CAMBRIDGE LIBRARY COLLECTION

Books of enduring scholarly value

Physical Sciences

From ancient times, humans have tried to understand the workings of the world around them. The roots of modern physical science go back to the very earliest mechanical devices such as levers and rollers, the mixing of paints and dyes, and the importance of the heavenly bodies in early religious observance and navigation. The physical sciences as we know them today began to emerge as independent academic subjects during the early modern period, in the work of Newton and other 'natural philosophers', and numerous sub-disciplines developed during the centuries that followed. This part of the Cambridge Library Collection is devoted to landmark publications in this area which will be of interest to historians of science concerned with individual scientists, particular discoveries, and advances in scientific method, or with the establishment and development of scientific institutions around the world.

Teneriffe, an Astronomer's Experiment

The experiences and challenges undertaken by C. Piazzi Smyth during his expedition to Tenerife on behalf of the Astronomer Royal are richly depicted and illustrated in this descriptive report of a major scientific expedition's course. The experiment was 'to ascertain how far astronomical observation can be improved, by eliminating the lower third part of the atmosphere'. This account of the data collection process details the flexibility and adjustments that were required throughout the course of this experiment, and the practical organisational difficulties and delights of leading such an expedition. The joys and interest of travelling in a foreign land are described with anecdotes of the people, flora, fauna and geography; particularly the research area, a volcano. Although first published in 1858 this detailed account of the experience of collecting precise scientific data in a challenging environment provides fascinating insights for any scientist undertaking research in the wild.

Cambridge University Press has long been a pioneer in the reissuing of out-of-print titles from its own backlist, producing digital reprints of books that are still sought after by scholars and students but could not be reprinted economically using traditional technology. The Cambridge Library Collection extends this activity to a wider range of books which are still of importance to researchers and professionals, either for the source material they contain, or as landmarks in the history of their academic discipline.

Drawing from the world-renowned collections in the Cambridge University Library, and guided by the advice of experts in each subject area, Cambridge University Press is using state-of-the-art scanning machines in its own Printing House to capture the content of each book selected for inclusion. The files are processed to give a consistently clear, crisp image, and the books finished to the high quality standard for which the Press is recognised around the world. The latest print-on-demand technology ensures that the books will remain available indefinitely, and that orders for single or multiple copies can quickly be supplied.

The Cambridge Library Collection will bring back to life books of enduring scholarly value (including out-of-copyright works originally issued by other publishers) across a wide range of disciplines in the humanities and social sciences and in science and technology.

Teneriffe,
an Astronomer's
Experiment

Or, Specialities of
a Residence Above the Clouds

CHARLES PIAZZI SMYTH

CAMBRIDGE
UNIVERSITY PRESS

CAMBRIDGE UNIVERSITY PRESS

Cambridge, New York, Melbourne, Madrid, Cape Town, Singapore,
São Paolo, Delhi, Dubai, Tokyo

Published in the United States of America by Cambridge University Press, New York

www.cambridge.org
Information on this title: www.cambridge.org/9781108014120

This edition first published 1858
This digitally printed version 2010

ISBN 978-1-108-01412-0 Paperback

TENERIFFE,

AN ASTRONOMER'S EXPERIMENT.

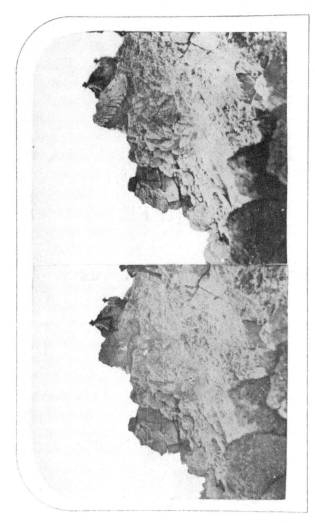

CULMINATING POINT OF THE PEAK OF TENERIFFE, 12,198 FEET HIGH, SHOWING THE INTERIOR
OF THE TERMINAL CRATER OF THE MOUNTAIN

p. 76

*Printed by A. J. Melhuish, under the superintendence of James Glaisher, Esqr F.R.S.
and published by Lovell Reeve.*

TENERIFFE,

AN ASTRONOMER'S EXPERIMENT:

OR,

SPECIALITIES OF A RESIDENCE ABOVE THE CLOUDS.

BY

C. PIAZZI SMYTH, F.R.S.S. L. & E., F.R.A.S.

CORRESPONDING MEMBER OF THE ACADEMIES OF SCIENCE
IN MUNICH AND PALERMO;
PROFESSOR OF PRACTICAL ASTRONOMY IN THE UNIVERSITY OF EDINBURGH,
AND HER MAJESTY'S ASTRONOMER FOR SCOTLAND.

Illustrated with Photo-Stereographs.

LONDON:

LOVELL REEVE, 5, HENRIETTA STREET,

COVENT GARDEN.

1858.

" . . . . . . they (telescopes) cannot be so formed as to take away
"that confusion of Rays which arises from the Tremors of the
"Atmosphere. The only Remedy is a most serene and quiet Air,
"such as may perhaps be found on the tops of the highest Moun-
"tains above the grosser Clouds."—NEWTON'S OPTICS, 1730.

TO

THE RIGHT HON. SIR CHARLES WOOD, BART., M.P.,

First Lord of the Admiralty,

WHO, BY HIS ENLIGHTENED PATRONAGE,

AT ONCE CONVERTED INTO AN ACTUAL AND SUCCESSFUL FACT,

A THEORETICAL IDEA,

LONG THOUGHT WELL OF,

BUT NEVER PREVIOUSLY CARRIED INTO PRACTICE,

THIS BOOK,

RECORDING THE CIRCUMSTANCES OF THE EXPERIMENT,

Is Respectfully Dedicated,

BY HIS OBEDIENT SERVANT,

C. PIAZZI SMYTH.

EDINBURGH.
December, 1857.

PREFACE.

IN the month of May, 1856, H. M. Lords Commissioners of the Admiralty, advised by the Astronomer Royal, were pleased to entrust me with a scientific mission to the Peak of Teneriffe. Their Lordships most liberally placed 500*l.* at my disposal for defraying the necessary expenses; and left me, within bounds of such expenditure, as untrammelled by detailed instructions, as any explorer could desire.

No sooner was the authorization known, than numerous and valuable instruments were kindly proffered by many friends of astronomy; and one of these gentlemen, Robert Stephenson, M.P.,—who had indeed fully appreciated the scientific question in 1855, and even asked me to accompany him to the Canaries in that year,—immediately offered the use of his yacht "Titania;" and by this, greatly ensured the prosperity of the undertaking.

The object mainly proposed, was, to ascertain how far astronomical observation can be improved, by eliminating the lower third part of the atmosphere. For the accomplishment of this purpose, an equatorial telescope and other apparatus were conveyed in the yacht to Teneriffe, in June and July 1856. There—with the approval of the Spanish authorities, (always ready in that island to favour the pursuits of scientific men of any and every country), the instruments were carried up the volcanic flanks of the mountain, to vertical heights of 8900, and 10,700 feet, and were observed with during two months.

On my return from this service in October, I had the honour of presenting to Government a short report on what had been done; following it, in the spring with copies of the original observations, as well as the results deduced. These were afterwards communicated by authority to, and read before, the Royal Society on the 2nd of June, 1857; when they were proposed for printing in the Philosophical Transactions.

Being then asked by various friends to prepare some account of the personal experiences under which the said observations were made, as likely to subserve many purposes not reached by the numerical statements of the Memoir,—I have endeavoured, in the following pages, to throw together those parts of my journal which seemed best calculated to bring out the specialities of scientific life, on a high southern mountain. Readers who would study the history, statistics, or physics, of Teneriffe, will find them treated of at length in the several admirable publications by George Glas, Viera, Von Buch, MacGregor, and Barker-Webb cum Berthelot; here, I have only attempted an humble record of particular labours, with due regard to the objects for which they were undertaken.

These objects, I am happy to say, have been so warmly appreciated by my intelligent and scientific publisher, Mr. Lovell Reeve—that although the book threatened to be very costly, by reason of the nature of the illustrative plates, he was prompt in relieving me of every attendant risk and expense.

THE ILLUSTRATIONS.

Anxious as myself to put all the actual facts of
Nature in the elevated regions that were visited, as
completely as possible before the public, Mr. Lovell
Reeve has been earnestly at work for some time past,
and with the gratuitous and continued assistance of
Mr. Glaisher of the Greenwich Observatory, has
succeeded in maturing plans for illustrating the
letter-press with a series of photo-stereographs, the
original negatives of which were taken by myself.

This method of book illustration never having been
attempted before, may excuse a word on this part of
the subject. By its necessary faithfulness, a photograph of any sort must keep a salutary check on the
pencil or long-bow of the traveller; but it is not perfect; it may be tampered with, and may suffer from
accidental faults of the material. These, which might
sometimes produce a great alteration of meaning in
important parts of a view, may, however, be eliminated, when, as here, we have two distinct portraits
of each object.

Correctness is thus secured; and then if we wish

to enjoy the effects either of solidity or of distance, effects which are the cynosures of all the great painters, we have only to combine the two photographs stereoscopically, and those bewitching qualities are produced.

Stereographs have not hitherto been bound up, as plates, in a volume; yet that will be found a most convenient way of keeping them, not incompatible with the use of the ordinary stereoscope, open below and well adapted for Mr. Reeve's new form of the instrument,—*The Book Stereoscope,*—constructed by Messrs. Negretti and Zambra, to fold up in a case like a map, without detriment to its stereoscopic action.

I have only further to observe, that while Mr. Reeve has been organizing his application of the manufacturing principle to the printing of photographs—Mr. Glaisher has personally superintended the chemical part of the process, in the hands of Mr. Melhuish, of Blackheath, in order to ensure permanence in the pictures so multiplied.

Edinburgh, January 1st, 1858.

ERRATA.

Page 37, last line,—*insert* "as" at end of line.
,, 54, fifth line,—*for* "70," *read* "40."
,, 64, last line,—*for* "strong," *read* "stony."
,, 65, second line,—*for* "a thousand," *read* "five hundred."
,, 94, last line,—*for* "fee," *read* "feet."
,, 155, sixteenth line,—*for* "lunologists," *read* "selenologists.'

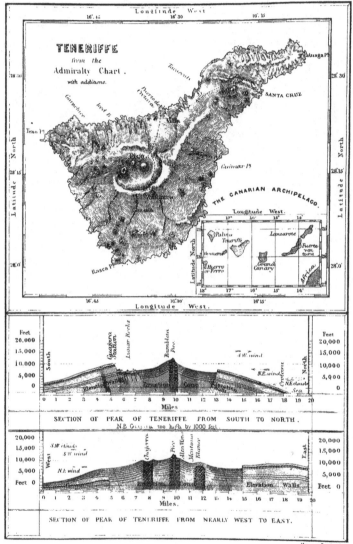

TENERIFFE
from the
Admiralty Chart.
with additions.

Longitude West.

Longitude West.

THE CANARIAN ARCHIPELAGO.

SANTA CRUZ

Anaga Pt

Teno Pt

Guimar Pt

Rasca Pt

Longitude West.

Palma
Tenerife
Gomera
Hierro
or Ferro

Lanzarote
Fuerte
ventura
Grand
Canary

Africa.

Feet
20,000
15,000
10,000
5,000
0

Feet
20,000
15,000
10,000
5,000
0

South

North

Guajara
Station

Lunar Rocks

Rambleta
Peak

S.W. wind

S.E. wind

Orotava

N.E. cloud

Sea

Mountain Cone

Miles.

SECTION OF PEAK OF TENERIFFE FROM SOUTH TO NORTH.
N.E. Crater too high by 1000 feet.

20,000
15,000
10,000
5,000
Feet 0

20,000
15,000
10,000
5,000
Feet 0

West

East

S.W. clouds
S.W. wind

N.E. wind

Chaorra

Peak

Alta Vista
Montana
Blanca

Elevation Walls

Miles.

SECTION OF PEAK OF TENERIFFE FROM NEARLY WEST TO EAST.

Vincent Brooks Lith.

CONTENTS.

PART I.

THE VOYAGE AND THE CLIMB.

PART II.

ON THE CRATER OF ELEVATION.

xiv CONTENTS.

PART III.

ON THE CRATER OF ERUPTION.

PART IV.

LOWLANDS OF TENERIFFE.

LIST OF PHOTO-STEREOGRAPHS.

MAP.

PART I.

THE VOYAGE AND THE CLIMB.

CHAPTER I.

SAILING IN THE TRADES.

ONCE within the parallels of the Trade-wind, every other natural phenomenon is found to give way before that grand commotion of the atmosphere. Therein does nature constantly seek—according to Lieutenant Maury's happy generalization—to restore to the air that moisture which was forcibly abstracted by the cold of the Polar circle; in order to be enabled, on passing into the opposite hemisphere, to distribute genial showers. The region of these winds appears eminently one of mechanical energy; for, driven by all the tropical power of the sun, its "cumuloni" clouds are ever hurrying along overhead, while the swell of the ocean hastens after them below; growing as it goes, and curling into foam, agitating and suffering agitation in a continually increasing ratio.

So we found it on board Mr. Stephenson's good yacht Titania, as she winged her speedy way to

Teneriffe, with her important freight of astronomical instruments, in the latter end of June and beginning of July, 1856.

Her beautifully-formed iron hull bounded over the waves as lightly as a pleasure skiff, and with far greater velocity. But how she did roll! in manner not disagreeably,—for a delicate digestion was not disturbed,—yet that the quantity was most notable, let twenty measured angles, of 15° to 20° each, per minute, sufficiently testify.

This rolling, be it understood, was but an expression of the disturbed state of the sea at the time; and, to a certain extent, would have been a necessary expression for any vessel in the same latitude and longitude; though each, according to its size and rig, would have a scale of its own. The Titania's scale we came to understand before long; and found its variations minutely sensitive to all the water changes, which the sailors described technically—as, long roll, cross swell, or head sea. We had now indeed, at last reached a part of the ocean, where the waves were gradually settling into a more regular and equable movement. But they ran higher than ever, were crested with foam, and gave the horizon a serrated look. Every now and then too, a billow, greater than his fellows, rushed by,—in the manner of the giant rejoicing to run his course,—and presently broke in a broad white sheet, like snow sparkling in the sun.

On such a surface, and under such circumstances, there was little display of animal life. Animals, as well as men, appear only to flourish with a certain degree of quiet. In the zone of calms between the southern and northern Trades, I had been struck, during former voyages, with the abundant manifestations of organic existence. So now, when every living thing seemed to have vanished from the face of the ocean, —we remembered that it was on the 3rd and 4th of July, or immediately before entering the Trade-wind region, that five whales had been seen sporting around the yacht, and that we had then passed through an enormous shoal of medusæ.

The least diameter we could assign to the collection was from thirty to forty miles; and at the rate of one to every ten square feet of surface,—which seemed to be a very moderate estimate, there must have been some 225,000,000 of them, even in the surface stratum.

Probably there might well have been so many to furnish food for their giant destroyers; but for numbers of victims, the whales were infinitely outdone by the medusæ. They were in the form of hollow gelatinous lobes, arranged in groups of five or nine, with an orange vein down the centre of each lobe. They moved slowly by sucking in water at one end of the lobe, and expelling it at the other—a principle of locomotion which, from time to time, inventors have proposed for steam-boats; but in the medusæ was attended, doubtless, with the means of straining the water of all its diatomaceous particles. Examining in the microscope a portion of one of the orange veins, apparently the stomach of the creature, it was found to be extraordinarily rich in diatomes; and of the most bizarre forms, as stars, Maltese crosses, embossed circles, semicircles, and spirals. The whole stomach could hardly have contained less than seven hundred thousand; and when we multiply these by the number of lobes, and then by the number of groups, we shall have some idea of the countless millions of diatomes that go to make a feast for the medusæ; some of the softest things in the world thus confounding and devouring the hardest, the flinty-shelled diatomaceæ.

When still off the coasts of France and Portugal, in the earlier part of our voyage, something new or strange, was diversifying every hour. The sailors most

heartily entering into our wishes, improvised nets,
rigged out buckets at the end of boat-hooks, and with
these and other ingenious pieces of machinery, were
very successful in fishing up curious things. Amongst
others, they caught certain swimming crabs, with
paddles in the place of claws; paddles—as they appeared, under a high magnifying power, of a rich
and tawny yellow colour, with a pattern resembling
bunches of hair-brown coral strewed over them, and
a blue-grey thread wandering here and there. While
we were still speculating on the use to decorative art
that might be subserved by this microscopic adorn-
ment, its fairy painting became confused and nearly
obliterated before our eyes, by the death and incipient
decomposition of the creature.

From the sea we turned to the sky, for in these
regions the forms of the clouds were so varied as to be
objects of never-ending interest; and the yacht's deck
was a famous place for meteorological observation.
The delicate cirri, and not much denser cirrostrati,
after exhibiting all their phases of beauty, passed at
length gradually, and before our eyes, into a peculiar
cloud, which may be called cirro-cumulo-stratus. It
is a long, thin, cirrostratus, high in the air, and
having a growth of compact little knobs and towers,

and, as it were, trees of vapour on its upper surface.
Thus garnished, it is a cloud often mistaken for distant land, is eminent for its picturesque qualities, and
indicates a strong tendency, or an approach to electric
discharge.

Some action of this sort was evidently going on in
the atmosphere; and amongst the other clouds, there
presently began to form the portentous " lightning-
cumuli," remarkable, even at mid-day for the flesh-
tint of their lights, and the steely-blue of their
shadows. After sunset, their small-featured, but dense
and grandiose masses appeared in dark relief against
the western sky. At the same moment in the east,—
these electric clouds being the only ones there, able to
reflect back any of the faint illumination thrown on
them from the opposite quarter,—they seemed to
glare out of the darkened heavens, like some pale
spectre of momentous size; and to shake their unearthly light over the waves, from the distant horizon
down to the very bows of the yacht.

At midnight followed grand displays of lightning;
awful had we been in the immediate neighbourhood,
for the forks were most vivid, and sometimes darted
in a group of seven or eight at once. At those
instants, so great was the blaze of light, as to reveal

a whole perspective world of cloud-land behind the flashes; sombre mountain ranges of cumulostrati, pierced with dark and long banks of cirri. The instant that the lightning gleam had gone, all this celestial diorama sunk down to a mere murky sky, with a suspicion of being streaked with clouds somewhat darker than itself. But the interest of the scene was still kept up by the ever-heaving and restless sea, which, breaking against the sides of the yacht, fell back in luminous foam, scintillating with stellar-looking points; and once the captain exclaimed, "there's a shark," as some large monster dashed on through the waves by our side, in a perfect path of light.

This used to be the state of things in that pleasant, and then somewhat idle region, to the west of the Iberian Peninsula. One of those days, nearly calm, was studiously spent by a big three-masted vessel on the horizon in forging its way over towards us. By six o'clock P.M. she had got upon our track, though considerably astern; and then rounding to, she showed her great broadside like a man-of-war commencing an engagement; but her intentions were peaceable. The huge Austrian merchant-ship, laden with corn from the Mediterranean, had toilsomely crossed the waters to the little English yacht, solely to ask the longitude !

The sunset was gorgeous that evening; bathed in golden light, and with massive banks of rich purple cloud, that, continually varying their outline, or rearing upon one end, and dropping jets of rain, clearly visible in the distance—appeared like huge saurian monsters contending in the air. A great change came on that night. The sky first grew heavy, then entirely overcast; a breeze from the N.E. began to spring up; hour by hour it improved; and, when the curtain of the next morning was drawn, everything told that the Trades were fairly entered, in the high latitude for them of 39°.

Meanwhile the clouds had broken, the old varieties were gone, and a new species, of very uniform character, was arranging its masses in long N.E. and S.W. cumuloni, and yet not cumulous, sort of striæ. The wind continued to freshen until it became a gale; but always from the same quarter; and always were the same long lines of cloud above us. They were evidently low, say 3000 to 5000 feet above the sea; coarse and puffy as to their character, and with no difference between their upper and lower outline. Half fog, half cumulus, they were pulled

out into thin long lines—lines interminably long—
for day after day we were sailing under them, and
never came to their termination. Ever were they
observed ranging N.E. and S.W.; and so parallel, and
yet so long, that their zones were seen morning,
noon, and night always radiating out of the N.E.,
and converging into the S.W., points of the horizon.
Nor did we cease to distinguish them thus, until
they finally coalesced into a general unbroken sheet
about the island of Teneriffe.

What, then, were these clouds, but a mechanical
effect and a sign of the Trade-wind; the cloud
material drawn out into streaks by the action of the
wind, just as it occasionally pulls into strings, the
foaming scum of the waves. Turner gained great
fame, in one of his paintings of a storm at sea, by
showing such parallel lines of froth along the whole
length of a great wave; but here were lines of
aerial foam that stretched from the coast of Portugal,
right away down the Atlantic, to the Canarian
Archipelago.

This sailing in the Trades—with one unvarying
form of cloud overhead, and that not very picturesque; and with a constant, everlasting blow of
wind, never rising to a storm, never sinking to a

breeze, without lightning, and without rain—is somewhat monotonous in itself, and makes an uninte-
resting time to many. But it was an opportunity
most precious to me, professionally, for trying some
not unimportant experiments in mechanics and
astronomy.

As thus :—

The general question of the angular motions of
a ship at sea, while curious to all, has an important
bearing with regard to nautical astronomy, the pal-
ladium of navigation. On shore, the astronomer is
always contriving the firmest foundations for his
telescopes, and even accuses stone walls of trembling.
How then is he straitened on board ship, when his
foundation there, does not vary only by two or three-
tenths of a second in the course of several months,
but by the amount of, it may be, 321° (three hundred
and twenty-one degrees) in the course of a minute ;
this being the number of rolls, multiplied by their
mean angular value, as actually found on board the
Titania by measurement.

Such an amount of movement must prove a complete bar, and in fact always has, to any sort of
observation, but such as can be carried out by Had-
ley's quadrant—an instrument in one respect inde-

pendent of a ship's motion. It is excellent so far as it goes, but still leaves many most important phenomena quite to the mercy of the waves—viz., any that require a high magnifying power, or that need a fixity of position in the telescope, and artificial referring-points for horizontality and verticality. Herein is included such a wide range of subjects, important to science, and useful to the sailor,—that remedies have continually been attempted from the days of Galileo down to our own, but as yet with utter want of success.

The fine old Italian fancied that he had discovered an easy method of determining longitudes at sea, by the simple observation of the eclipses of Jupiter's satellites; and he proceeded to bargain with the King of Spain, then the strongest naval power, for a magnificent reward, in return for so many telescopes to be made by himself. On shore his tubes proved perfectly equal to their work, that of showing the Jovian satellites clearly and distinctly; but, on a ship's deck, alas! no mortal man could use them.

Galileo's difficulty is still unsolved; an observing chair, or stand, on which a telescope may be placed, is still a desideratum for the sailor. Not a few such machines have been attempted in times past on the principle of a pendulum; and the inventor of one of them (Nairne), in the days of Captain Cook, was rewarded by Government. Prematurely; as he failed in practice; and for the reason Sir John Herschel has well pointed out, in the "Admiralty Manual for Scientific Inquiry;" viz., that "free suspension tends rather to perpetuate than to eliminate disturbances."

The dinner-tables on board the yacht, hung from pivots, were on a plan once proposed for observers, and answered wonderfully for mere dining purposes. Bottles and glasses remained steadily in their places, even when the tables seemed occasionally to swing so low, as to be pointing downwards nearly to the carpet under our feet. But it cannot be too distinctly understood, that keeping a dinner-service from falling, or its liquid contents from outpouring, is a perfectly different thing to preserving a platform truly level, as regards either the horizon or the direction of gravity. The latter is what is needed for astronomical observation; and that a swinging body does not procure it, even when a glass of water is kept from spilling, may be powerfully demonstrated, by attaching the said glass to a string held in the hand, and swinging it,—not only through large arcs, in imitation of the rolling of a ship,—but round and round, through the whole 360°. Let this be tried by any one, and no water will be spilt; the dinner requirement will have been accomplished; but, oh! for the astronomical one; as the surface of the water, during the experiment, will have successively deviated from the horizon through every possible angle.

With a clinometer of new construction, first tried during this voyage, we applied measure to the case; and found that when the deck inclined 18 times per minute, through an average angle of 15°,—the tables in question rolled 42 times in the minute, through about 7°. The tables were heavy, with their hanging weights (several hundred weight), and their pivots were endued with much friction; but still their oscillations, as pendulums, during the rolls of the ship were perfectly sensible. These, sometimes opposing, and sometimes combining therewith, tended greatly to increase the confusion of the movement. With the barometer, a small weight in lighter gymbals, the vibrating effect was still more prejudicial.

Hence, again, the propriety of Sir John Herschel's views, of trying to modify the evils of *free*, by somewhat *stiff* suspension. Yet the evils are thereby only modified, by no means removed; as every increase in the stiffness of the suspension, in proportion as it checks the oscillations of the table as a pendulum, must communicate the roll of the ship. The only complete and radical cure, then, is to be sought for in something altogether different from the much-tried pendulum principle.

Such means of rectification appear to exist in the force which keeps the axis of the earth in a constant position, as it annually revolves about the sun : viz., the rotation of a heavy body round a free axis. Models of an apparatus on this plan were shown at the French Exhibition, and are described in the "Transactions of the Royal Scottish Society of Arts" for the years 1855 and '56. Suffice it now to say, that on July 3rd, on board the yacht Titania, in lat. 41°. 16′. N., long. 10°. 51 . W., with the rolls of the vessel amounting to 25 per minute of 10° each, and the pitches to 27 per minute of 3° each, we tried the largest apparatus of this description yet manufactured.

The only available place on deck for the erection of the instrument, was at the vessel's extreme stern, with a look-out over the quarter; just where most up and down motion occurs, and with full effect of rolling. This, however, was well for testing to the uttermost the new principle, which had this feat set before it—to enable an observer, without using his hands, to keep a telescope constantly directed on a distant object in spite of the waves. A sort of observing box had been provided, which, while allowing insertion of the observer's head and hands, as well as affording a free opening for vision, kept out the rudeness of the wind. But the essence of the means to ensure the all-important steadiness, was a wheel one foot in diameter, eleven pounds in weight, suspended and balanced in gymbal rings; as well as capable of being put into rapid rotation by two trains of wheels, acting on either side of its axis.

The captain called up the sailors to drive those wheels. Two strong men at each handle, and the striking together of hundreds of steel and iron teeth many times in a second—as the axles spun round with a velocity almost unprecedented in practical mechanics —soon produced a thrilling sound that called up every one to see what was going on. They found it of course a most unnautical proceeding.

Presently, on the wheels being thrown out of gear, and the gymbal rings unclamped, the sound died away, though the revolver went on spinning. Then following with my head the small apparent motions of the eye-piece, I looked in, and had the

c

satisfaction of finding the horizon of the sea remaining steadily in the field of view. All the rolling of the vessel could avail nothing against the power of the free-revolver principle. Adjusting the balance, and then bringing the sea-line on the wire of the telescope,— it actually remained bisected for a considerable length of time; and the captain, the first and second mates, and many of the crew, were invited to look at it one after another.

They saw, and readily confessed the fact, of the useful thing that was now accomplished for the first time at sea; and throwing their prejudices behind them, they took kindly to the scientific inno-vation.

The sailors worked with enthusiasm. When-ever the driving handle moved in the direction for pulling, they fastened on a rope to it, and clapping their feet against the timbers of the yacht, pulled away as only sailors can pull. They pulled till the multiplying wheels, with their innumerable striking teeth, shrieked again in their velocity of rotation.

Some splendid spins were thus obtained, which gave to the table, with nothing visibly supporting it, a firmness like a rock. Touch it, then, incautiously—and it resisted like a wild beast; but press judiciously on the gymbal rings—and the table was adjusted more accurately than by any tangent screw. The action improved with every increase of speed which we could bring the wheel up to; and every additional trial saw it revolving a greater number of times in a second.

Annoyed only at not being able to get up the full velocity at once, and not understanding the mechanical difficulty of causing an eleven-pound wheel to revolve one hundred times in a second—as indeed few persons do, or have any idea of, without trying,—the willing hands put out more strength still. Then came a sudden crash, and in a moment the men lay flat along the deck; the strong steel driving axles, each an inch in diameter, had broken. So the observation of Jupiter's satellites, and sundry other intended crucial experiments, were deferred to a future voyage.

On July the 7th we had passed the parallel of Madeira, and were approaching the Canary group at the rate of nine or ten knots an hour. The forms of the waves, I marked in my journal as grand; the colour of the water, as a fine deep Prussian blue,—a change from the cobalt blue of the first part of the voyage; and their material as endued with an extraordinary power to froth and effervesce—for a bucket

drawn upon deck, went on sparkling for a long time
like champagne. A land bird, probably from the
Salvages, perched for a time on our rigging; and the
question was thereby prominently brought before us,
when shall we sight the Peak?

From how many persons have I heard of the 90
miles, or 100 miles, and even 140 miles from which they
had seen this lofty mountain; or of the two or three
days that they have been sailing past, ere it was finally
lost below the horizon! What a termination to a
voyage, the high cone of the Peak! What a land-
mark; how easy to make a port so defined! Such
are the theoretical ideas on the navigator's problem;
but how different the reality, as will presently be
seen.

The lines of cumuloni, which had long been growing broader and denser, overspread the whole sky in
the course of the day; and, with a perfectly cloudy
evening, a wild windy atmosphere, and huge boiling
waves all around, what hope for us to see the Peak
from any extraordinary distance? Nay, at night we
even began to be anxious about our position, it was
so long since we had got any observation; and here
we were still scudding along over the tops of the
waves, with a velocity equal to their own. On, how-

ever, and still on we went all the night through, in the direction concluded for Teneriffe, and surely by sunrise we must see the great Peak.

Before sunrise there was a report of land! On deck stood the captain closely scanning a range of rocks and breakers, some eight miles to the S.W. Every now and then, as the yacht rose on the now giant waves of the Trade-wind, we saw the surf beating on these rocks with terrific distinctness. But what their summits were like could not be told, being lost at a small height above the sea in the sheet of the Trade-wind cumuloni, that stooping low, enveloped them almost as a mantle of fog. What with the roaring gale, the densely-clouded sky, and its concomitant low-toned colours, as well as the cold watery look of the rocks and everything about them, one was inclined to ask, what had this at all to do with Teneriffe? The "Island of the Blessed," the site of the Elysian fields, according to poesy, the hot and sunburned African isle of travellers, beaconed with a cone 12,000 feet high, as a landmark for all wandering sailors! Yet this little fragment of misty-looking breaker-coast—a bit that could, without blame, be mistaken for innumerable parts of the Scottish or almost any coast—was all we had to

judge by, of our proximity to the burning mountain
of tropical Africa.

The captain believed the rocks to be at the N.E.
point of Teneriffe, and at nearly the distance, and in
the direction he wished them at that moment.
But with the frankness and honesty of a really able
seaman, he confessed that they might be on the N.E.
coast of Grand Canary. Currents, he said, which
sometimes do, and sometimes do not, run with great
violence in these parts towards the east,—wreck-
ing Atlantic vessels on the African coast, and giving
their crews as slaves to the Bedouins,—might have
been setting the yacht eastward since the last astronomical observation had been obtained. But if that
be so, he added, " by keeping on so many more miles
on our present course, we shall see a distant headland
come out immediately behind these rocks; while, if
it is really Point Anaga, in Teneriffe, the rocks will
gradually trend into a long-continued coast line."

An hour more, and this nice question in practical
navigation was settled. The captain was right; we
were rounding the north-eastern cape of Teneriffe.

We then edged in a little more towards land. As
the day advanced, the clouds rising in altitude displayed the steep slopes and cliffs of Anaga. Nearer

and nearer we passed in review one headland after
another; the colours next began to be distinguishable,
and very red, brown, and yellow colours they were.
A little longer, and having got somewhat under
the lee of the land, the force of both wind and sea
was immediately moderated. Now, with but gently
rolling motion, we began to perceive the terraced
garden-walls on the cultivated slopes; to admire the
wild grandeur of many a mountain glen opening down
toward the sea; and presently to make out the solitary
euphorbia bushes overhanging the rocks; while in the
distance appeared a long low line, which the telescope
soon split up into the houses and steeples of Santa
Cruz.

On the hill summits, the wind was playing
wild work with the clouds. Our old friends, the
cumuloni, were being dashed against the rocks, and
the tops of the ragged rocks; torn piecemeal, and then
hurried over the edge into dissolution; much after the
well-known scene of misty drapery on Table Mountain, at the Cape of Good Hope, during the prevalence
of a "black south-easter." With Teneriffe, however,
more in the heart of the Trade-wind, its clouds were
more abundant, while the extent of high country to
catch and bewilder them was greater, and in its form,

infinitely more picturesque. Suddenly a gap appeared amongst the driving masses, and lo! there was the Peak of Teyde whitening in the morning sun. In a few minutes the vista closed again; but it had crowned the truth of the captain's navigation, and at the same time left indelible traces on our minds of some of the characteristics of this great volcano; its long external slopes of gentle ascent, the sharp-pointed though obtuse-angled peak, and the light bright colour of its pumice-strewed soil.

This unveiling for a moment of the chief glory of the island, showing it for an instant as a reward after the toil of the voyage, and then shrouding it in mist and in mystery, as one advances nearer to the coast, —leaving it to faith and perseverance of the "excelsior" vein, on landing, to push through the entangling luxuries of a southern clime, and re-discover the great Peak in a new world above the clouds,— this momentary manifestation of a higher and purer sphere,—is by no means so accidental an affair as might be thought at first, and is therefore no unfrequent phenomenon to vessels entering these roads. It has been described by Darwin, in his "Naturalist's Voyage," with beautiful point; and appears to depend physically, on a certain line of separation between the land-cloud and the sea-cloud, which is more or less constant through half the year. Hence, conversely, for the same period there is given to observers on the Peak, raised high above these lower strata of mist, and looking down through their narrow partition, the vision of a certain strip of ocean, and that only, some five or six miles beyond the Santa Cruz anchorage; everywhere else, they see only broad plains of barren cloud, spread far and wide below them. The discovery of the reason was the result of subsequent experience. At the instant of the phenomenon appearing to us, the effect on the feelings was such, that there could have been few persons with whom the leading idea would have been the physical explanation.

CHAPTER II.

SANTA CRUZ.

WITH the wind still mangling the clouds on the serrated crests of Anaga, and ruling supreme over the dark and mist-covered sea to the N.E.; but, nevertheless, with an admirably warm and bright day about us, we rowed ashore through the numerous vessels in the bay. They were rolling to an extent, that from at first seeming dangerous, grew at last to be positively absurd; so intent did they seem to be on first giving us a peep down their hatchways, and then trying to show the state of preservation of their keels. This was the effect of the swell from the eternal Trade-wind outside, prodigious in amount for a harbour, or a place where ships lie at anchor; but being here glassy and harmless, we rowed pleasantly over it—now catching a sight of the shore-line, with white surf breaking along its rocks; and now seeing all apparently submerged under a wave close by, up to the tops of the highest steeples of the town.

VOLCANIC "BLOWING CONE" IN OROTAVA, ON THE NORTHERN COAST OF TENERIFFE.

p. 55.

Printed by A. J. Melhuish, under the superintendence of James Glaisher, Esq. F.R.S. and published by Lovell Reeve.

At the mole, what a scene!—what a place for my wife to land at! for there, though the structure is carried out into deep water, the swell is not so inno-cuous. Crowds of boats are about, and the place is alive with men, mules, and merchandize of import and export. Every few seconds comes a great wave, heav-ing up all the boats one after the other, and then letting them down crushing and grinding together; while the turbulent billow, rejoicing in the mischief it has done, rushes along in its appointed course, half deluging that side of the mole. Scarcely have things recovered from this visitation, when on comes another great wave rearing out of the vasty deep, and tries to outdo its fellow, or overtake it before reaching the line of sandy shore—where little boys, as brown as a berry, are bathing in noisy crowds. This sort of ocean game keeps on all day, and day after day, in this most open and exposed of roadsteads. How the island must rue the loss of its ancient port of Garachico, filled up by a red-hot lava stream in 1705.

Our boat was small and frail, but the Spaniards adroitly eased it on its way, as their surrounding launches rose and fell, swashing up and down with every surge. At length we almost touched the wall,

and seizing the instant of being on the point of a
wave—we stepped lightly ashore, in a land that told
abundantly, though the sea outside had not done so,
of a southern latitude and a tropical sun. The scene
that had suddenly burst on us,—who had been undergoing the dismal winter of 1855 and '56 in the British
Isles, and had had nothing but heavy rains up to the
last day of June,—was chromatically in another hemisphere. It would have been a paradise to a painter
from the raw and gloomy north,—colours so dazzlingly
rich, yet so harmoniously combined, and such ideal
forms met the eye on every side. Men and women
and children were there, of whom literal portraits
would be perfect pictures—rich, too, in the poetic
element.

The peculiar tint of the Spanish complexion is an
easy one to introduce and to harmonize amongst
other colours; witness the predilection of even landscape painters for brown trees, brown grass, brown
everything. How, too, the hue is set off here by the
white garments, glowing in the bright sunlight, and
the rich red scarf that the poorest porter wears about
his waist. Entering, in the course of the day, a
Scotch merchant's establishment in the city, we saw
a roll of the most gorgeous scarlet satin—the purple of

the Roman emperors—laid out before some peasants; poorly enough clad generally, but by no means disposed to forego indulgence in a piece of finery, manufactured perhaps in Glasgow or Macclesfield, but never there exposed to public gaze. How every painter, and eke every tee-totaller too, should thank the men who live on barley and water and silk-finery, in place of spending their means on rich food and strong drink, practically synonymous with riotous and unlovely living.

Carefully let us pick our way amongst the troops of loaded mules, and the crowds of scarfed men and hatted women, erect in their gait, and brilliant in their coloured garments. The matrons amongst them seem generally to wear a dark or scarlet shawl on their heads, with a black hat above. This shawl is allowed to droop in graceful folds down the back, and the young damsels similarly display a white or yellow kerchief, but are more commonly seen without the hat. The "head" drapery, indeed, pendant behind, would appear to be a necessary adjunct of female dress in Teneriffe—no doubt because in this burning climate, it protects the spinal marrow of the wearers from the hot and piercing rays of the sun.

With all these distracting novelties, so particu-

larly interesting after a long voyage, let us be
wary that we impale ourselves not on the horns of the
oxen that tranquilly, rather than lazily, wend their
way along through the crowds of porters, and drag
behind them liliputian sledges, with box or barrel
placed thereon. What classical models of symmetry
are these little oxen; from top to toe they are all
of one fine tawny colour. None of those clown-like
piebald marks that badge the domestic animals of our
Saxon country, preventing a sculptor from fully perceiving the play of the muscles, no such rude blotches,
appear in these exquisitely natural-looking creatures.
In uniformity of colour, and that a tint greatly to be
admired, they have all the lordly air of unenslaved
denizens of the forest; uniting therewith a tender
and honest expression in the full liquid dark eye and
pendant eyelids, which so took the fancy of the
Greeks.

A camel, that presently comes swaying along with
a grand piano slung on one side, and a heavy bag of
sugar to balance it on the other, appears rather out
of his element. So he is, too, for though this eastern
end of the island, that looks towards Africa, and in
the parallel of the Great Desert be it remembered, is
hotter and drier than the western portion—it is yet far

from reaching the tension of the continental Sahara. We have here light and heat in perfection, but happily for man, and his comforts, some little moisture also.

Hence, when walking at mid-day in one of the basalt-paved streets, each glittering stone sending back the full rays of a vertical sun, and the gleaming houses on either side affording a steady white-hot glare of unmitigated sunshine—what words in a northern language can express the delightful emotions, when at the open gateway of one of the semi-Moorish abodes we look in upon a grove of bananas! Throwing a tender green shade over the interior court, their grand and delicately-structured leaves rise up aloft, catch the fierce rays of the sun before they can do mischief, receive them into their substance, make them give out the most varied yellow greens; pass them on from leaf to leaf subdued and softened; pass them on to the oleander's fountain of rose-pink flowers, to the dark green of the orange, the myrtle, and the bay; and leave just light enough at last in the green cavern below—to show the bubbling of some tiny fountain, the welling heart of this fairy oasis. Our fashionables who visit Italy and Spain in winter only, how little do they know of the province of the sun.

Save us, however, from too long a continuance of
it. For when we saw English gentlemen domiciled
in the Canaries going about like young girls with
green silk umbrellas over their heads, and defending
their eyes with great blue goggles, that must have
marred to the wearers all pictorial characteristics
in every landscape,—when we found them so list-
less at home as to let flies walk uninterfered with
about their noses while conversing with us,—we felt
thankful, indeed, for our present stamina. Hardened
in the cold of British winters, we found now, for a
time at least, in lat. 28°, that we still enjoyed, in
spite of an almost tropical sun, the ability to roam
about as freely as in England; studying all the
aspects of nature, as well in her scenes of bright
magnificence, as in her secluded realms of gloom and
grandeur.

So high an authority as Humboldt has stated
(" Personal Narrative," Bohn's edition, vol. i., p. 48),
" That from every traveller beginning the narrative
of his adventures by a description of Madeira and
Teneriffe, there remains now scarce anything un-
told respecting the topography of the little towns
of Funchal, Santa Cruz, Laguna, and Orotava."

Coming half a century after the subject was thus pro-
claimed to be utterly threadbare, I did not expect
that anything still remained, which could be usefully
noted by a casual visitor. Yet, when that great
traveller himself goes on to say of the " little" town
of Santa Cruz, " On a narrow and sandy beach,
houses of dazzling whiteness, with flat roofs, and windows without glass, are built close against a wall of
black, perpendicular rock devoid of vegetation," that,
" reverberating the heat, unduly increases the temperature;" and when we found, on the contrary, that in
the Santa Cruz of 1856 are lordly mansions with glazed
windows, and ascertained it to be a large town, with
gardens interspersed; separated too from the beach,
which lies altogether in front of it, and at a lower level;
when we found the town to spread backward from
the shore over an extensive, slightly inclined plain,
that stretching away at the same gentle angle of not
more than five degrees, and for several miles from
the sea,—is nowhere terminated and backed by a
" black, perpendicular wall of rock,"—other ideas
came over us. We could not then but suspect, that
Humboldt's description applies so little to the present
time, either as to social features or natural topography, that there is room enough still for another

wanderer, and perhaps many another, to add his mite towards elucidating the very interesting characteristic features of the capital of Teneriffe.

In search, however, for the "black, perpendicular cliffs," we strolled through the town, away from the sea; saluted an occasional palm-tree, that soul-moving emblem of the East and the South; and discussed the merits and promises of pumpkins and fig-trees, in those gardens where we could look over the stone walls that bounded them. We wandered away till the gardens of the town, and the villas of its citizens, passed into the cactus plantations of outside farmers; and seeing these gradually rising in the distance into the terraced lands of genuine country peasants—we agreed to leave further exploration in that direction, until we should be driving there, as drive we did two days after, with four horses at full speed, on our way to Laguna and Orotava. So we turned again, and perambulated the town in other directions. We examined the shops, and stood astonished before one filled with bamboo instruments, the like of which we had never seen before in all our born days; yet surely we had dreamed of them, or were dreaming then. The olive-cheeked damsel, with blue-black hair and dark-beaded eyes, in charge of these mysteries, immediately volunteered an explanation; and, with a little clever pantomime, quickly demonstrated, that the cabalistic implements were nothing but "distaffs," and were in general use amongst her countrywomen.

In the churches we could not admire their pictures; but the "Correggiosity" of the Spaniards in decoration was not a little remarkable. The greater part of the building was kept in gloom, deepened by sombre colours; while towards the altar there was one vast blaze of gilded ornamentation.

Not vulgarly laid on in flat, meaningless surface, was this profusion of the precious metal; but, wreathing and twining, it appeared like an inextricable mass of tropical foliage; yet subdued, methodised, and conventionalised with infinite tact, to permit an orderly and harmonious effect in conjunction with architecture.

Was it from having made the whole region about the altar, from floor to roof, a mass of red burnished gold, that the rearers of this temple had exhausted their scale of glorification; and, to distinguish anything more glorious still, must take a lower tint on their pallet? or, was it that the remembrance of Mexican mines and ores of Peru have left an ecstatic halo in the Spanish mind about the idea of silver? We know not, but so the fact was, that we found the altar composed of silver, while the chapel, nay, the whole end of the church, was of gold.

Not so much did we approve the woodwork of buildings, whether sacred or profane; interesting reminiscences though it gave of the old arabesque. Certainly descended from the stock of the Alhambra, are all the window and door frames of Santa Cruz de Tenerife; but at this distance of time, the style, sadly debased, reflects on its parentage; hinting that it was but curious ornamentation, or luxurious trifling, and not high art at all. That which was the carver's work formerly, or at least was concealed by, and glorified in, the art of the carver, is now but a sedulously-wasted piece of joinery. A door that might be made in a single panel, in two, or in four, shows here in fifty or more panelled frames. No angles or proportions of beauty seem to have incited, only the idea of intricacy and multiplicity. A fish-slice, M. Gautier would say, is nothing to it; and all that has been gained, is, an infirmly-trussed bit of carpentry, with innumerable joints.

Had the material been still, as of old with the Moors, the knotted root-stocks of the olive, this lili-

putian panelling would have been highly commendable; but when the material of all Santa Cruz
carpenters is now, the broad and honest plank of
American deal, what need is there to fritter its
length away, in panels only three or four inches
square?

Other features of the buildings, however, called forth
much admiration. For, comparing the general architecture of this town of the Spaniards, with Cape Town
in nearly equal southern latitude, and not far from
the same longitude; and where two Saxon nations
have followed each other, reproducing in a hot and
thirsty African land,—one the marsh-built structures
of Amsterdam, the other, houses appropriated to a
region of snow and mist,—we could not fail, despite
its unmechanical carpentry, in giving the palm to
Santa Cruz.

The Moorish arrangement of a dwelling appears
pre-eminent in suitability to a burning climate; for
it affords in the interior court, such charming retreats,
for man and choice plants from dusty winds and
the blinding light outside. Nowhere else than in the
shade and shelter of a whole environing house, in
these lands swept by the constant Trade-wind, could
such masterpieces of vegetation be produced,

certain that we saw, when treading the interior gal-
leries of verandahs, in many of these delightful mansions.

How different though, and how primitive, are
the residences of some of the lower classes, under
sheets of lava, which form a substratum for all the
country around; and after passing under the town, are
broken down just as they enter the sea. Of but a few
feet in thickness, these beds expose at this point the
clay which they formerly overflowed, and often caused
to bear crimson testimony to the once glowing heat of
their melted stone. Now, sea spray acting on the
whole, washes out the soft material here and there,
and forms a long row of caverns. Roofed by projecting sheets of basaltic lava, they enjoy in front
a magnificent prospect of bright blue sea, the
shipping in the bay, and distant purple peaks of
the picturesque island of Grand Canary. In his
own hot and dry climate, with his all-defending cloak
to wrap himself in, the Spaniard cares little about
the artificial *agrémens* of his sleeping-room; so no
wonder, if in this island, where the climate is warmer
and drier still, these lava-covered halls are numerously
tenanted.

Wending our way next along rocky descents towards the sea, the ice-plant (*mesembryanthemum crys-tallinum*) with its succulent leaves, frosted with ruby and diamond-like drops, appeared in abundance. Con-triving to grow on mere heaps of rubbish, the inha-bitants find the succulent creeper a cheap and easy means of manufacturing soda. That throwing this plant into the fire, should turn it into stone, is, say the more thinking among them, one of the greatest wonders in the world.

Near the mole a net was being drawn ashore, and, as more and more of it was brought to view, a sudden storm of vociferation and malediction arose. The fishermen had caught sight of their enemy, a great ray (*Pteroplatea Canariensis*), with a sting like a stiletto. The huge square, flat mass began to flounder in the utmost agitation, as if anticipating its fate; and was no sooner on *terra firma*, than men and boys of all ages rushed at it with sticks and oars, and im-plements of all sorts; using them in such style, that one could soon examine the creature's dreaded weapon of offence with perfect *sang froid*. Hammer-headed dog-fish were numerous; and being taken out care-fully one by one, and having had their heads well knocked against a stone, were thrown into a heap to go to some oil-preparing works; a large assortment

of eatable fish, between mackerel and herrings, still
remained behind.

Enough, however, of mere gazing about a strange
town. The Peak, with which our professional work
is connected, is staring at us all this time; some might
say it is rather peeping; for, seen from Santa Cruz,
only the upper fortieth part in height of this 12,000
foot mountain, a portion merely of its miniature terminal cone, rises above sloping hills to the west of
the town. There is just enough seen to testify to,
and no more ; and the position is curiously undigni-
fied for so mighty a volcano. Often one has difficulty
in distinguishing it from little parasitic craters, red
and yellow, seated on the intervening ground; and
such small affairs, that one might think them the
work of recent navvies. On visiting them, they come
out grandly enough ; but in this land, distance does
not so much beautify far off things in the clear, transparent air, as make them appear close, and be thought
small. The Peak is visible from here, certainly, but
that is all; it is thirty miles away, and cannot be considered as belonging to this end of the island. The
English Consul and merchants are most obliging, as far
as lies in their power, but that does not include the

Peak. For the ascent of it, they can give no better advice, than that we should betake ourselves to the town of Orotava, close under the highest part of the volcano, and from there arrange our mountain operations.

Santa Cruz, the capital for government and for commerce, has at least answered our purpose for reporting ourselves to the Spanish authorities, and requesting their countenance to our mission. With their accustomed liberality in this island, its dignitaries have made us as free to wander wherever we like, as the natives themselves. Orotava, they say, we may visit when or how we will, by sea or by land, to stay as long or as short a time as we please.

On this permission we act at once; yet, before we can leave our hospitable friends, the sun has set. Darkness, indeed, is descending over land and sea—and the clouds on the craggy hills are accumulating again, after their partial rout by the heat of day—as we row back from the mole once more to our floating home. The swell comes rolling in grandly from the open ocean. For a moment the dark line of a wave, seen high against the sky, threatens to submerge us; and then up we are mounted on its back to view the red fires of fishermen, in their Neapolitan-looking

boats, similarly rising and falling, and throwing lurid
reflections among the billows. The yacht is descried
at last; and there it is rolling away from side to side
as actively as for a wager; as if it had not had the
luxury of a really good roll all day long; or as if it did
not know that it was going to have full opportunity
for the enjoyment every day, for at least three months
to come. I knew it, however, and considered the subject anxiously, for the captain was to make metereo-
logical observations on board, coincidently with the
mountain party aloft; and the angular motions of a
vessel at sea, are capable, we had found, of sensibly
influencing the height of a barometer, in a manner not
hitherto taken into account.

CHAPTER III.

A T 4 P.M., on July the 10th, my wife and I rode into Orotava. We knew that we were now comparatively close to the Peak, but it was concealed from view by strata of mist, descending so low as 3000 feet, and extending over the greater part of the sky. The air was nearly calm; for the N.E. Trade-wind, the tyrant of Canarian seas, cannot blow home to this huge mountain, that, unlike the lower crests of Anaga, towers high above the influence of polar currents.

The streets were trafficless, and quiet reigned everywhere. The only noise, beyond the clattering of our horses' feet on the pavement as we passed along, was the pushing open of a trap in each wooden window, to allow of the protrusion of some curious head; and then the falling to again of the little panel when curiosity was satisfied. Our two attendants, the Spanish grooms, running alongside the

horses, after having, in pride of their city, taken us
up and down as many of the streets as possible—concluded by a grand career through the principal square,
and a spirited entry into the courtyard of the inn,
awakening all the echoes of the place. This could not
be the inn! but " Si, senor," they answered, it was
the inn, the hotel, the grand hotel, and that whereat
we were expected; and away they vanished with
our tired steeds.

We seated ourselves in a corner of the court; we
noted the vesicular basalt wherewith it was paved;
we observed the several stories of verandahs surrounding the enclosure; the strong featured furrows of its
red-tiled roof gleaming against the sky, and projecting far over the walls; while at one corner a belvedere
tower rose quaintly, and not unpicturesquely, above
the whole. All this time not a sound was heard, not
a form was seen to move, save a chance butterfly
tumbling in by accident over the housetops, and
hastening out again as fast as it might. By what
name can we call this silent inn—this inn of the
dead—this most dead and death-like inn? One began
to realize prematurely the feelings of the last man.
How needlessly, for there all the time, concealed by
the darkness of his cool retirement, is the young pro-

prietor, placidly smoking his cigar, and contemplating us, never imagining that he has anything more to do.

In half an hour we were visited by the acting Vice-Consul. The journals of all ascenders of the Peak invariably bear thankful testimony to the abundant services and effective agency of the consular authorities in Orotava; and we were to experience the same assistance on a yet more extensive scale. Mr. Goodall, the present officer, proved exactly the man we desired; of few words, he listened to others, observed for himself, disappeared with his cigar; and, before it was burnt out, had arranged his plans, and begun to put them into execution.

The reception of the yacht was the first thing considered. So next morning, when her flowing sails and raking masts were seen in the offing, an experienced pilot was ready to go on board and bring her to the anchorage. What an anchorage, though! The water was inconveniently deep, and not 200 yards off was a black beach bristling with rocks, and overhung with precipices; while every moment the great surges rolled in from the ocean, and broke in a perfect avalanche of foam. Such a position on the British coast would have been dangerous indeed. Who could

tell, but what a wind would not spring up before the
morning, and from a quarter most devoutly unde-
sirable. That may be in England, said the Islenos;
but here, in Teneriffe, our wind in the summer season
is no affair of chance, and you are as safe as in a
closed dock. Now, too, they added, you are anchored
fast, and have no wind at all, so you can't help your-
self, and must remain where you are ; no wind, either,
will you have until to-morrow morning at sunrise,
and then a nice little breeze from the S.W. will
enable you to put to sea in gallant style.

So the captain philosophically made the best he
could of it; and while instruments and packages
were being handed over one side into large native
boats — on the other side the urban aristocracy
were crowding on board to see his far-famed yacht.
I do not know that they were quite competent to
appreciate the beauties of her build, but they admired
the splendour of her internal decorations mightily,
and stood in mute astonishment before a fire-place
in the main cabin. This was a puzzle, and perhaps
not altogether a fair one.

The inhabitant of a city without chimneys is not
bound even to expect a fire-place on board ship; and
that which obtained in the yacht, one must confess,

with its white marble and inlaid porcelain, had not much of a Vulcanian look, in spite of the iron bars and the brass wire screen. Hence, when a knot of the most learned had examined, and discussed, and concluded that the thing must be a " bird-cage,"—if they did not make a very fair first approximation,— they certainly produced a useful hint on meaning and appropriateness in decorative ornament, for many of our manufacturers at home.

With the instruments and camp equipage, were landed the carpenter and the second mate of the yacht, William Neal and William Corke. They were obligingly picked out from the whole crew by the captain, following strictly Mr. Stephenson's liberal instructions, as likely to be the most serviceable men ; and they entered heartily into the unknown work before them. In a marvellously short time they had packed up their hammocks and necessaries; and pre- sently stood on the beach of Orotava in the blue uniform of the Royal Yacht Club, the observed of all observers.

These lookers-on were many indeed, being half the children and all the idlers of the town, who from this moment began to dog our steps wherever we went, notwithstanding some interference from the alcalde

in our behalf. The maimed and halt, too, trooped
down to the inn; and, when kept forcibly out of its
court, the poor creatures would seat themselves perseveringly in the grand square opposite; and whenever we looked out of the windows, it was to behold
a waving of diseased arms and leprous legs.

In a large, unoccupied billiard-room we arranged our
numerous packages; and were happy to add to our
councils Mr. Charles Smith, a former graduate of St.
John's College, Cambridge; resident in Orotava for
twenty years; and while keeping up his mathematical
lore, adding practical science thereto, and, above everything else, a specialty for all that appertains to ascents
of the Peak.

Long and earnestly we discussed the best methods
of carrying out the objects of our expedition. The
main purpose thereof—to ascertain the degree of improvement in astronomical vision with the elevation of
the observer—appeared to demand, that two or more
stations should be occupied at different heights, but
was indifferent to the full extent of the sky being
visible at each place. Other, and not unimportant
objects of research, imperatively required full vision
of the whole hemisphere of the sky, down to, and

below, the horizontal line. The difficulty of compassing all these requirements, was not a little augmented by the amount of mere locomotion, that would be necessary on the sides of a mountain—of whose vastness and ruggedness, I was assured over and over again, that I could form no idea without actually ascending. One person even insisted that there could be no living at all at a great height up its flanks, because the heat of the sun was there so terrific by day, and the cold so intense at night. But he being soon answered, that such effects were in themselves an infallible proof of the greater lustre with which the stars would be found to shine, the cherished end of our journey to ascertain, and to measure,—we recontinued the course of our discussion.

No station in or about Orotava, nor within 4,000 feet of altitude from the sea level, could be thought of for a moment; for the local cloud of the Trade-wind was now established in permanence for the summer; and rendered all the country below its level as cloudy and as untoward for any astronomical observations, as the very climate of Scotland itself. Again, the summit of the Peak could not be occupied; for not only was it unattainable to mules,—by reason of the ruggedness of its rocky slopes,—but the abundant escape

E

of hot vapours, from the highest crater of this yet
unextinct volcano, would be quite enough, we agreed,
if tried at the place, to spoil any accuracy in telescopic
vision. On the sides of the Peak, a station practicable
so far as related to the conveyance of instruments,
might be found : but no one could say whether there
might not be escapes of volcanic breath in the neighbourhood; and all were certain, that the great cone
towering above, would cut off a large portion of sky,
fatally for many subsidiary observations with which
we were charged.

What high point is there then, at a distance
from the Peak, and from all influences of Plutonic
heat, and commanding a full view around on every
side? " Guajara," shouted Mr. Smith; " that is the
mountain for you : it is 10,000 feet high, four miles
to the south of the Peak ; and, excepting that, is the
highest land in the island." The maps were con-
sulted. " Are you sure Guajara is 10,000 feet high?"
" No, not certain, but it has the credit of being so;
and even if, as is usual in such cases, a fraction of its
height should be removed by instrumental measurement, I know well that it overtops all other mountains in the Island, but the Peak itself. There too,"
he added, " high above all the clouds, you will have

an intensely blue sky overhead, a dry air around, and
a spring of water not a quarter of an hour's walk
below you; by no means an unimportant conside-
ration even for an astronomer."

Then Guajara for ever! and now let us see about
the baggage. Here are tents, camp furniture, and hut
building materials, with carpenters' and masons' tools;
meteorological instruments, physical apparatus, and of
photographic chemicals not a few : there are also the
Sheepshanks telescope, and the great Pattinson
Equatorial. Hola, poco à poco! you don't think of
ever getting up those awful big boxes of that mighty
Equatorial! And why not? Why not, indeed, but
you don't know the mountain! The distances are too
great for men, and no mule could carry one of those
chests. Then two mules can carry it between them,
as in the Sicilian *Lettiga*. Such a thing was never seen
in Teneriffe; and the mules here are so decidedly more
than proverbially obstinate and vicious,—that it is as
much as one can do to get them along in the good
old-fashioned ways, without any attempt to impose an
innovation upon them. They would not submit to it
any how; and besides that, the road is often, nay
continually, so narrow and so crooked, that there
would not be room enough for two mules and the

box, on shafts between them, to turn the corners.
Oh, you don't know the mountain!

Then, if mules cannot convey the boxes, they must
be taken up by relays of men. Well, perhaps sixty
or one hundred men might do it in several days, if
we can prevail on the hardiest men of Icod el Alto to
come; and if you level and broaden the road for them
all the way; but as it is now they could not carry the
boxes at all; they must employ poles, and if you
place them cross-ways, the roads are not broad
enough; if length-ways, they cannot turn the corners.
Only listen to what the best *portadóres* will say:
here are some who have just crossed over the central
ridge of our island from the south coast, and report
that there is now such fine, dry, and bright weather
up above the clouds, while we are living down here in
gloom and in steam. The manly bronzed muleteers
were brought in, and on being told that the big boxes
were to be taken up the mountain, appeared sure that
their contents must be something as light as bonnets;
but on half-a-dozen of them trying to lift one, they
groaned and vociferated and stamped at the very
idea; the " Caballero," they said, could know nothing
about the mountain.

Certainly I knew nothing of it in the way that all

the peasants of the island did, and that was the important way for present purposes. The sooner then, that I acquired such knowledge the better; for thereby alone could I stand on an equal and respected footing with those to be employed on the service. The distances were too great to permit valuable time to be wasted in going backwards and forwards empty-handed, and that was not the way to prove the qualities of a mountain-road for heavy transport. On Monday, therefore, I determined to start with all the people and all the boxes that mules could carry, and the work on Guajara should then be commenced. So the huge Equatorial chests were definitively pushed on one side, while all the other and more compendious packages were arranged in a single stratum on the floor; in order that the different carriers, whom Mr. Goodall was rapidly engaging, might come and feel the weights, choose the loads for which they would be responsible; and cogitate on the manner in which they would make them fast, on the backs of their unhappy mules. The number of animals required, was quite unprecedented for journeys up the Peak; but the acting Vice-Consul assuring us that they would be all forthcoming on Monday morning, and that he himself would be there to see—we made a truce with anxieties, and strolled through the town.

Not fifty yards had we traversed over the basaltic paved streets, when lo! at the end of one of them appeared a " volcanic blowing cone." Some 70 feet high, it towered above the surrounding houses, and exhibited a beautiful parabolic figure, like the Hindoo temple of Bindrabund on the Jumna, so admirably pictured by the younger Daniell. Composed of hard lava, and with an upper aperture still yawning, whence the burning breath of fires below once issued in fury and with destruction,—and may do so again in spite of the huge cross planted on its summit,—what a history this " fumarole" revealed. And when several similar cones appeared rising amongst the houses in various directions, one could not but recognise an eminently characteristic feature of the town, thus built, like Herculaneum of old, not only at the foot of a great volcano; but exactly in the line of its shortest road to the sea, into which its lavas have poured, and will certainly pour again.

We must have a photograph of this cone; so back we go for the camera, and returning with it ready charged with a collodion plate; we present, fire,—and

immediately have secured the cone and the distant
hills capped by cloud and terraced with gardens on
their flanks; while besides the Spanish cottages and
the basalt pavement in the foreground, we have a
curious damsel peeping out of a hanging trap-door
in one of the wooden windows. (*See Photo-stereograph,
No. 2.*) Again we return to the charge; and on another
plate—public curiosity being now more fully awakened
—we have not only the inorganic elements of the scene,
but three children's heads peering over each other, like
the knobs on a mulberry; a father, son, and two daugh-
ters standing in their doorway, and a timid young
woman, with a white handkerchief thrown gracefully
over her head, peeping round the corner of the street.

A third plate in the camera; we right about face,
and take another of the volcanic blowing cones,
and a whole street filled with wondering spectators.
In developing this picture, we were much struck by
the appearance of one of the houses, rising in height
above the rest. The whole front face of it seemed to
be made of vertical bars, and the wings on either side
being similarly furnished, one was reminded of the
dens for the one big and the two little elephants in
Wombwell's menagerie. Some wild beast must be
confined there, was the sentiment as we gazed through

the wet plate at the exquisite perspective of vertical
bars, a subject that photography so loves to reproduce.
We revisited the place, and found indeed a very good
den for giraffes and possibly for elephants; but the
only occupants were three large crosses.

Very famous crosses they are; the patron crosses
of the city. Orotava was in early days named
after one of them, "The Port of the Holy Cross;"
but the success of one brought the others into the
field, with great confusion of the *fêtes* which were in-
stituted and are still held in honour of each. Attempts
are being made to amalgamate these holidays of the
different crosses; but with more than vested rights to
combat, the reformers have made no sensible progress
as yet.

Large wooden crosses are affixed here to private
houses, as well as public buildings; and every now
and then in the streets, they meet you face to face.
There is a huge one nailed on our inn, and at this
moment a stealthy cat is using it, as a means of
climbing into the window of an upper story. What
a cat! and what creatures seem all the Spanish cats!
The dogs, mangy and famished, are bad enough; but
the cats,—oh! may we be preserved from them. A
type of the whole class attends our meals; its hair

dishevelled, its skin hanging loosely about its bones;
with its tail cut short, and its ears cropped, it looks
at you as though it were contemplating to get at your
very life-blood. You give it food, and it rushes
at the morsel franticly and unthankfully. At the
same instant its kittens rush too, all mangy and
unkempt, from corners where they had lain concealed;
then begins a most disreputable scene, mother and
children clapperclawing each other, mewing and
howling and tearing the food out of each other's
mouths. The serving boy comes in to change a
course, and disperses the ravening crew by running at
them with a jug of water; but the moment he has
left the room, back they gallop furiously from their
hiding-places, and plunge headlong promiscuously into
a dish, which the youth, not being Briareus-handed,
could not carry away with the plates.

After a disturbed night, Sunday comes on, calm and
quiet and genially warm. Every now and then a
gentle breath of air rustles the leaves of broad poplars
that border the grand square. A fine palm tree in the
distance, waves slowly its long magnificent fronds;
while ever and anon, borne on the wings of the south
wind, comes floating a melancholy yet musical sort of
jangle, of richly-toned church bells. Throughout the

day we can just hear the deep far off booming of the
surf, on its rock-bound coast; and still, ever and anon,
comes that most melancholy yet musical, tangled
jangle of bells. Those Spanish church bells! one
cannot fancy them pulled by man's methodical hand;
they must be hung on the branches of some broad-
spreading tree, and swept all together by the
passing of a sudden wind, like the plaintive strings
of an Æolian harp.

The sun is about to set. Do our ears deceive us?
No! it is a military band, and there come the soldiers,
a very French-looking edition of soldiers indeed. They
enter the Grand Square, march round its open area,
and draw up in a circle at one end. Fashionables, like evening moths, begin to flutter around the
centre of attraction. Gentlemen in the latest cut of
English habiliments appear, and walk by themselves.
Then come ladies, also by themselves, two and two,
with dark tresses, black eyes, and the gauzy graceful folds of the deservedly immortal mantilla. They
whisper their pretty sayings only to each other, though
exhibiting the inimitable working of their fans to
everybody. Little young ladies also come out, dressed
in the most absurd degree of the British boarding-
school miss; and what with their tiny bonnets on the

back of their heads, their short petticoats sticking out
behind, and their laced-up silk boots, one cannot fancy
them descended from those ideal beings so aristocra-
tically veiled in the mantilla.

The band plays, the promenaders promenade perseveringly, and the poorer children also congregate on
the square. Round-headed things, of a bronze complexion, and lightly clad, they seem to enjoy themselves infinitely more than their young betters, stuck
out so preposterously with the expensive fashions of
another land. Not the least joyful too of the whole,
are a couple of coffee-eyed imps who dance unabashed
before the whole assembly, in a dress consisting of
nothing but portions of a brown shirt, that barely
hang piecemeal about them.

Day is closing, the band plays its concluding piece,
with much the feeling of " God save the Queen," and
is marched off the ground. The gentlemen then venture to approach nearer to the ladies of their respective
acquaintance; in the dusk of the evening, daring to
offer them an arm, and to see them safe home. The
streets grow silent, night comes on apace, and before
nine o'clock has arrived—we see nought but the stars
that shine tranquilly down, on the Grand Square of
Puerto de Orotava.

CHAPTER IV.

BEGIN THE ASCENT.

WITH more than twenty horses and mules at daybreak on the 14th of July, clattering on the
flinty pavement, and with as many men shouting to
their respective animals, and disputing among themselves for the lightest loads—there was no more sleep
from that early hour, for any one in the inn.

Well did the muleteers understand the art of
making fast on pack-saddles a large variety of goods.
We distributed rope liberally amongst them, they
twining it ingeniously over the burdens and around
the bodies of the creatures; then taking a bend
of it under the belly, they inserted a tough curved
stick, and twisted ferociously, until every turn of the
rope became perfectly taut; or the mule, as it falsely
seemed, lifted almost bodily off his four legs, would
have shrieked if he could, at the vehement stricture of
all his internal organs.

Such of the men as had completed these arrangements, wanted to start off at once, and we had much difficulty in restraining them. Certain ideas, were in our mind, of keeping them altogether in the ascent of the mountain, overlooking their proceedings, and making a lightly loaded mule occasionally change burdens with one more heavily laden. We should not perhaps have made such strenuous attempts as we did towards this end, had we been fully aware of the utter impossibility of putting it in force throughout the whole day's journey; or had we been as well informed, as the experience of three months subsequently made us, of the honesty and perfect trustworthiness of these carriers. With merely a general impression that they are to make the best of their way over the long rugged heights, they start off, in the ordinary course of traffic in the island, with an overwhelming load of heterogeneous and uncounted goods; and there is seldom a complaint of the loss or non-delivery of any of the items.

Five horses led by five men now trotted up, for the sailors, ourselves, and Mr. Andrew Carpenter, a nephew of the acting Vice-Consul, who was to act as our interpreter. Before mounting, we looked closely to the supply of water; so many barrels for the men on the journey; and so many for ourselves at night, in case

we should not find the reputed spring on Guajara,
or in the event of its proving dry. This being a mid-
summer precaution of Africa by no means to be
trifled with. Under dreadful penalties too, was the
man in charge of our water mule, warned to keep
always close in front of us; for it would be no local
act of dishonesty at all, were any of the thirsty
guides, when once on the journey, to get the animal
into a quiet corner, and drink out of the tabooed
casks to his heart's desire.

The long procession, and such queer-shaped packages as most of ours were, drew many an early-rising
spectator about us to behold. Mr. Goodall and Mr.
Smith were there to assist. The latter mounted his
horse and accompanied us some distance from the
town. The ascent begins at once, in the very street,
nay, at the sea-beach, or possibly far under the
surface of the sea; for everything indicates that this
whole island of Teneriffe is, even in its entirety, but
the summit of a half-risen mountain. The whole
surface of the country ascends at an angle of from
5° to 12° for several miles in distance; and though
this be but a moderate angle indeed for the side of a
mountain, it is so unconscionable for inhabited lands,
that winter rains are threatening every year, to

PEAK OF TENERIFFE FROM OROTAVA, ON THE NORTHERN COAST.

P. 403

Printed by W. M. Ebrard, under the superintendence of James Glaisher, Esq. F.R.S.

and published by Lovell Reeve.

wash away all the softer soil. Only by paving every pathway, and intersecting every garden with walls and ramparts of solid stone, can the descending waters of a short but heavy rainy season, be prevented from cutting grooves from top to bottom of the mountain. (*See Photo-stereograph, No. 3.*)

Presently we came to an open stony tract. " Here," said Mr. Smith, " before 1829, lay some of the most fertile and charming vineyards of Teneriffe; but they were all carried away in the great flood, or rather cataract, that rushed from the mountain on the night of November 6, of that year." A lamentable scene of desolation it appeared to us, and in an island that could ill afford to lose any of its hardly won territory. A long description is to be found in the first volume of a great work on the " Canaries," by Barker-Webb, and Berthelot; yet we must regret, as the storm was one of unique force and anomalous character, that observations of the strength, direction, and variations of the wind; heights of the barometer, and the quantity of rain in different parts of the island,—did not take the place of passages so little instructive in physical research,—as, " at the sight of this terrific disaster, in the memories of the good inhabitants of the village, I interrogated Providence,

and demanded of Him, if this destiny was not an
error."

We are still ascending; there is little wind in the
early morning air, the sea is heavily, the mountain
and sky above our heads are but thinly, clouded;
while low slanting beams of the sun, striking
through mist openings in the East, lend a variety
together with a vague magnificence to the long perspective of cultivated slopes. These are backed by
rocky ridges, gleaming here and there with bright
lines of lava dykes, and pointed in many places with
the so frequently double head of a volcanic crater.
Now, as we bring up the rear, we can see the whole
of our lengthy cavalcade, like some huge serpent,
wreathing along upwards and downwards and from
side to side, as the lesser irregularities of the ground
demand.   Crimson silk scarfs and white garments
of Spaniards come out brilliantly from the yellow-
brown of the soil; and somehow or other, our biggest
sailor, mounted aloft on his horse, seems always going
over just the highest part of an intervening ridge,
in the deep purple-blue of his Royal Yacht Squadron
guernsey, looming powerfully against the mist of
the morning.

Between garden-walls, by a narrow and strong

pathway, we wound along, and passed on our right
an interesting volcano, about a thousand feet high;
and presently another, almost the same size, on our left.
From a distance, we understood them well; and saw
in each, the red mound, with a vast hollow at the top,
breaking through its surrounding wall on one side,
towards the sea; and thence pouring a long stream
of thick, viscid material, that hardens as it flows.
But on coming close, the real vastness of their size
prevented our seeing more at once, than so moderate a
portion, as not to be very distinguishable, by form
alone, from any Neptunian hills. Yet these two volcanoes are but mere small warts, upon the flanks of
the monster we are scaling; and it has many such
excrescences, some much larger, others smaller.

Perhaps this gradation of scale has led certain
geologists, thinking of the chain of parasites, and
those two immortal lines,

> " Great fleas have little fleas upon their backs to bite 'em,
> And little fleas have lesser fleas and so ad infinitum;"

to call such features, " parasitic" craters; and by that
one name to fix what they are, and thereby, though
not intentionally, to stop further inquiry into their
real nature.

F

If the name could do that, there would be much in a name; but some few persons who will not be so guided, demand, that besides the existence of a scale from great, to little, and to lesser, as well as the position of one on the top of the other,—there be also proved to obtain the remaining characteristic quality of parasites, viz., that of small ones living on the substance of large ones; or, as the poet expresses it, "upon their backs to bite 'em." This has not been done, and, most probably, cannot be done ; because, according to much that has been ascertained of volcanic action, all the several crater vents have their origin by ducts from one and the same central cause; and in place of a parasitic nature, have rather the aspect of the many heads of a hydra, or the crowd of snakes all springing equally from the blood of Medusa, not from each other.

By half-past seven we had attained the perpendicular elevation of 1800 feet, and still the only road was a narrow rough-paved footpath, bounded by solid stone walls, or by natural ramparts of rock. Surfaces of this material sometimes took the place of artificial pavement. Rude as such they were, though not more so than to be expected on a mountain side; and they hardly justified the frightful account of a gallant

French traveller, concluding with, "that if a cava-
lier was to fall from his horse on these sharp-edged
rocks, he would break both his arms and his legs."
The inequalities of this road would be made nothing
of by a Cape waggon,—the angle of ascent would
even be smiled at; but where would there be room for
the wheels, to say nothing of the long train of oxen?
" Give me room to drive," I was inclined to sigh in
Archimedian strain, "and in one of those South African
vehicles I will easily carry up all the much-dreaded
boxes of friend Pattinson's Equatorial." But where
was there room enough? Not in the narrow pathway
we were pursuing, without pulling down the side walls
all the way along, and rounding many a sharp angle;
and not in the open country on either hand, without having to cut through artificial banks of earth
and stones, to an impracticable extent, at every
few yards.

We gazed in despair at this so-called open country;
as far as we could see in every direction, it appeared
like a transverse section of a gigantic honeycomb, so
intricate was the interlacing of its horticultural
system of fortification, against the sweeping vengeance
of winter torrents. Here were new experiences indeed; the soil, the climate, the plants were African;

but this universal mural cultivation, which enclosed every few paces of ground, was something entirely opposed to the agricultural system of any great continent, spreading out unlimited extents of un-cared-for wilderness, before its colonial hunters and shepherds.

So much were we taken up with these ideas of the shop, as not to remark the botanical changes which were going on; until suddenly we found, at a height of 1900 feet, that the gardens on either side, in place of oranges, lemons, figs, and peaches, were now chiefly filled with pear-trees. Two thousand feet, and lovely wild plants of the hypericum, in full and abundant bloom, with their delicate young pink leaves and rich yellow flowers, were intruding in every corner,—2400 feet, and a few heaths were caught sight of; 2800 feet, and English grasses began to appear.

We turn about at 2900 feet, and behold! we are even with the clouds; which, but scanty this morning, disperse in our immediate neighbourhood, when we seem just about to enter them. Several miles off to seaward, however, there is the ominous front of a stratum, some thousand feet higher, and of immense thickness. Its massy rollers surge up one behind the other, and threaten to break in upon the land. But for all that, they do not on the whole, sensibly pass their boundary. A long pull now followed under the hot sun, blazing in a sky of unbroken blue; and by nine o'clock, reaching a hollow at the height of 3900 feet, the whole party halted, to rest the animals and take breakfast; with the ocean of white clouds far below us, and concealing all the lower country from our view.

In a ravine on the left, we found a delicious shadow under a part of its Eastern bank, somewhat steeper than the rest. Large heaths, ferns, and occasional laurels waved immediately before us, while a passing flock of milch goats furnished all that was necessary to complete our meal, and to produce even exhilaration under such health-inspiring circumstances. A horse, being much longer than a man, in devouring his food, we had time enough to examine into some curiosities of the long, strange chasm, whose cool shade we were enjoying. Much had we noted its appearance during many parts of our ascent in the last 1000 feet. Deep and broad and terrifically wild it was, with its blue basalt rocks, here worn smooth, and there broken and tossed about in chaotic masses.

Were there any traces of ice-action? would ask an Agassian geologist. Well, frequently there were groovings visible, but only in the centre of the ravine; and on a space of rock not more than two feet broad.

Quite close to our impromptu breakfast-parlour, lay one of these grooved channels; it appeared even a good specimen of glacier polishing down and scratching, and some of the contained crystals were cut clean through, level with their matrix. A portion of the surface so marked was rent by many cracks, and I tried to knock out an apparently isolated bit; but in its adamantine strength, it laughed at the efforts of my puny hammer.

The chasm had been but of moderate size, until the terrible deluge of 1829, when in a few hours it became a broad gulf frightful to see. In some places the depth was increased, as well as the width; and there, we have no groovings. The principal effects were sideways, and by means apparently of undermining the strata of rock, which is here wholly composed of basaltic streams of lava, lying one over the other. Much of the material thus loosened was carried down the mountain, and spread over the vineyards below; other portions, for the water broke up and prepared, far more than it could carry away, were left behind

almost *in situ*. These latter are still to be seen, like
the Titanic stones half-finished in the ancient quar-
ries of Baalbek; for no subsequent flood has reached
their level; and in the dry atmosphere of this height,
above the lower clouds, there is little decay. Monstrous blocks we thus found, rough, angular, and with
all their natural cleavage as sharp, as on that awful
night, when they were forced up from their beds in
the wild hurly-burly of the headlong waters.

Speculations are often made on the ages in geology,
based on the length of time, that water must require
to cut through a channel of given depth, when the
material is of extreme hardness, and the observed rate
of increase, something nearly insensible. On these
grounds some myriads of years would have been required, to cut out the hollow where we breakfasted.
But Nature does not by a red-tape routine, restrict
herself to one only mode of working, but rather varies
her plans, as she finds most effective in each case;
wearing down a soft rock, and breaking up a hard
one. Such had been eminently the case here, where
one night in the natural method, had done the work
of ages on the theoretical. Such was the case also
on the sea-coast to the west of Orotava, during a storm
of last winter. There, in two or three days, and per-

haps chiefly in a single tide, the sea,—urged on the
land by an impetuous N.W. wind,—rose to an unprecedented height, washed for several hours into the
Grand Square of the town, and carried away part of
the neighbouring coast.

On visiting the scene of destruction, and removal,
for that was also well effected,—we found a fine
bay, eating some distance into the land, and bordered by magnificent basaltic cliffs, much like the
Salisbury crags of Edinburgh.   They had this advantage perhaps, that they were quite new, and perfectly
un-weatherworn, for their material was as hard blue
whinstone as may be found anywhere.   Whole centuries will make no sensible progress in filing them
down, or trying to decay them away.

Water then may evidently, from these Teneriffe
experiences, cut out as extensive valleys in hard rock
as ice; and that without acting, as some contend,
solely as a carrier of hard, rasping materials.   But
the physiognomy of these two classes of excavations
will be very different; roughness and angularity, as
our refreshment ravine showed, being produced by
the breaking up and washing out of water; smoothness and general rounding off, by the action of ice.

CHAPTER V.

SOON after 10 o'clock, A.M., we were again mean-
dering in long broken line on our upward way, with
the clouds below, and a brilliant sun shining above.
It was splendid climbing; a mountain ascent made
very easy, was this riding up the gentle slope. Here
we were, already at a height considerably above the
top of Table Mountain,—to compare one of the
islands with the continent, of Africa,—whose vertical
precipices begin at half its height. But on Teneriffe,
for upwards of 6000 feet, are still no greater average angles than 12° to contend with; and in most
places so much soft soil, that after a shower of rain,
there would be little difficulty in turning furrows with
the plough, over a considerable part of the surface.

At 10h. 50m. we had reached a height of 4700
feet, and the first specimen was met with, of an interesting leguminous plant, to which we were afterwards to be greatly obliged, the "*codeso*" of the

natives; the "*adenocarpus frankenoides*" of botanists.
With closely packed composite leaves of light and
warm green, a yellow flower, woody stem, branches
like a miniature cedar tree, and with the bark of ages
hanging about it, this specimen of the "*legumineuses
frutescentes*" of the French savants, bore a certain acting resemblance to the "doorn booms," or thorny
acacias of South Africa, whose place it appeared to
supply.

At 11h. 10m. the height of 5280 feet was attained,
and a solitary pine-tree was seen, the last unhappy
member, at this spot, of forests which once girdled
the mountain. Some heaths and a few ferns were
also observed, but the aromatic *codeso* chiefly occupied
the zone. The summit of the Peak was now detected
gazing at us through the sunlit air, hazy with intense illumination; and at 11h. 50m. our anxious
looks were rewarded by a specimen of the "*retama*,"
(*cytisus nubigenus*), that unique mountain broom, the
like of which none of the other Canary Islands, nor
any of the African isles, and in fact not another spot
in the world can show. Spanish broom grows in the
gardens below, and, with Scottish broom, will grow
almost anywhere amongst the habitations of men, if
seeds only be sown; but he who would behold the

retama, must ascend more than a mile vertically into the air.

We were travelling now over pretty rough ground, immediately along the edge of a deep gorge, displaying in section several cataracts of lava. On a portion of the opposite side was a steep slope of volcanic rubbish, loose, cindery, and one would think in such constant motion that no plants could retain a footing thereon. Nor can any of them whatever do so, except this admirable *retama*, and that rejoices in the site, and flourishes. How wonderful the adaptations of Nature to the necessities of different regions. For here, where the ceaseless motion of the sliding particles composing a hillside, destroys every other living thing; where the aridity of the soil during many months is only surpassed by the aridity of the air, which is drier than that of Sahara,—Nature has produced a plant, that on the mere remembrance of winter rain, long since evaporated, can furnish no contemptible supply of wood; and with its richly-stored white flowers, arranged in close rows along its smaller branches, affords illimitable honey-making materials to all the bees of the country.

At 6560 feet, the lessened slope indicated that we were entering the circle of the Canadas. We could not

see very far around, for the surface was rough, and we
were winding along in single file by narrow sinuous
paths, amongst rocky hillocks, the results of confused lava streams. Here the interstices began to be
strewed with a new material, a light sponge-coloured
pumice-stone ; while alternating bushes of *codeso* and
retama, at considerable distances from each other,
formed the sole vegetation.

At 0h. 45m. P.M., and at a height of 7150 feet, a
gust of south wind in our faces, together with less undulation of surface, and a full appearance of the Peak
now rising up grandly on our right, left no doubt
that the ascent of the northern slope was finished ; and
that we were travelling over the basin of the ancient
crater, a crater whose vast dimensions (eight miles
in diameter) can hardly be paralleled save in the moon
itself.

Here, the surfaces of pumice-stone soil, widened
out ; the rocks, red and jagged, became fewer ; the
codeso disappeared ; high land was seen on our left, and
presently,—as we entered on quite an African-looking
desert of white sand and yellow stones,—a fine range
of blue mountains was seen to the south-east and
south. What mountains are they, what can they
be ? Why, they are merely the opposite sides of this

gigantic crater we are crossing; and the highest of
them and one of the most distant, is that " Guajara "
on whose summit we hope to arrive before night.

Long and depressing is now our weary way, stretching over to the eastern side of the crater, across its
pumice-strewed floor. One o'clock, P.M., and our
elevation is not improved; 7127 feet, says the sympiesometer, or 23 feet lower than our last observation.
Hence mountain climbing is over for a while, and
with it have evaporated much of the exhilaration and
energy, which the presence of distinct objects to be
overcome, invariably excites. We have now transcended all the strata of clouds, and have entered a
most moon-like region. The flaming sun set in the
middle of the sky above our heads, within seven
degrees of the astronomical zenith, showers down his
merciless rays, through the thin transparent air, on
every side. The sympiesometer acquires a temperature of 100° as it hangs on my back. Light and
heat revel everywhere; there is no need of volcanic
assistance.

In spite of orders, in spite of vociferations, the
line of mules and men will become broken; divers of
them will keep falling behind; the binding-ropes

of their burdens become slack in the drying air, and must, they say, be tightened; while somehow the water-mule is always with the stragglers. Men are caught and denounced in the fact of drinking at our barrels, but claim them as theirs, by virtue of some changing of loads on the journey. A drum-head court of inquiry is instantly held; the expeditionary barrels, intended for Guajara, are recognised beyond a doubt. With their mule they are made to journey ever after, between two of the riding horses; and if anything goes wrong with the pack-saddle, we wait and assist personally at its rectification. A small supply of water I have all the time, in a double tin-box under lock and key; but that will not last the party long, if other sources should fail; and not a drop of moisture have we seen yet, in all this dry and weary land, since we left Orotava in the morning.

Two o'clock, and we are pacing under the eastern wall of the great crater. First come the sides of steep hills of loose brown burnt stones, in a state of slow but continued descent. In every one thousand square yards, or perhaps more widely separated still, is a *retama* bush, but not a particle—not a glimpse of any other specimen of vegetation. Undisturbed by man, this region shows you the history of whole generations of the plant :—First, the little tender seedling, yielding in the direction of its growth to the slow grinding avalanche of clinkery minerals; underneath which, however, it has contrived to establish a root of most precocious length. Then comes the ambitious young bush in youthful vigour, curving back its stem to regain ground that was lost in its infancy, and vertically over the root, sending up to the blue vault of heaven, a rich fret-work of filamentous shooting branches. Next see the full-grown plant, with a hemispherical mass of flowers throwing a grateful shade over the thirsty soil; and with dignity maintaining its original place, while all the surface of the hill, by little and little, goes crumbling and scraping and sliding down past it.

Time passes over its boughs; one by one they drop off, and as dead limbs immediately begin to share in the downward motion of the outer particles of the hill. Last scene of all, a short stump of root-stem is left, to point out where the bush once displayed its beauty, while all its old component members, are found further and further below; and decayed, in proportion to the number of years that have elapsed, since their vital principle fled.

The gradual dissolution of the numerous twigs, the

progressive disappearance of all but the mass of harder
wood towards the centre of the thickest stems of a
dead branch, and the degree of bleaching, followed by
reduction to fibre of that part, seemed to give a very
tolerable measure of the comparative length of time
that such weathering had been going on. So we compared the general state of decay of old branches, with
their distance from the parent stocks, and obtained
unexpectedly coincident rates of descent for the hills
of rapilli; very slow certainly, but not a little sure
and regular.

These cindery ridges gradually passed, as we
travelled on, into horizontally stratified cliffs, the
strata admirably distinct for miles; sometimes many
hundred feet in thickness, sometimes thin as ribbons;
while here and there a dislocation and discoloration
occurred, as they were traversed by dykes, proceeding
in radial lines from the Peak, the centre of the basin.
One remarkable dyke particularly drew our attention,
exhibiting as it did magnificent blades or plates of
greenstone, rising some thirty to forty feet above the
soil on either side, and stretching from bottom to top
of the crater-wall. Occasionally I left my steed, and
climbed the slopes on foot, to break off and secure
specimens of the most characteristic rocks, all felspa-

thic; but found, that with so large a cavalcade, and
under the anxious circumstances of a first ascent for
ultimate astronomical purposes, I could do but little
in this way.

The real business of the expedition being astronomical; the mules with their unaccustomed loads of
scientific instruments had to be looked to circum-
spectly; if too far out of reach, they would lie down
and roll; if too close, they would kick out at you or
at each other. Fatigue and heat, with hunger and
thirst, were beginning to tell on both quadrupeds
and bipeds. But on, and on, all must go; for this
crater plain is no place to bivouac in.

Cold enough at night,—stormy beyond measure in
winter,—this elevated region, by day and in summer,
exhibits to a wayfarer only the desolation of light;
but that is more than sufficient to parch the tongue, to
pain the eye, to give the full idea of a desert. For us,
the air is filled with scorching solar rays; from the
sand, from the rocks, from the Peak some three miles
on our right, and from the cliffs so close on our
left, the light and heat rebound. In this glaring
atmosphere, as slowly and heavily we labour along,
and with shaded eyes peer into the volcanic perspective,—now is seen some desolating flood of lava,—now

a gleaming plain of cindery dust of pumice; here
level as a lake, there teeming with thousand groups
of reddened slags, like so many chemical preparations, simmering in the mighty sand-bath of the
mountain's laboratory. Next, in hot mid-distance,
a low eruption crater rears its dark and fissured
head over the wilderness of lava crags between.
Then anon you pass under a group of monstrous
chimney apertures; that gape, all together, from a
red protruding mass hanging threateningly over
your path; and lead you to imagine that here, but
lately, the expiring forces of the once too dreadful
volcano, had united themselves for one more effort;
and at this very spot, had vomited forth for the last
time their fiery breath. These are the invariable
elements before us, in this glittering desert of light:
where a happy green plant, a leaf, or a blade, is
found,—never.

As we proceeded, the perspective of the small, on
the floor of the large, crater continually altered. In one
position it showed a single cone, in the top of which
there would only be room for the smallest pit; but a
little further on, that appearance was discovered to be
merely the effect of the pointed, highest part of its
wall, overtopping the rest; for presently the general

shape of its whole mass, was that of a broad double-
headed cone. These heads continually widened as we
progressed, until they were at length separated by
more than their own height; and then the whole
interior of the hill lay open before us, with a
breach which might be easily entered on the eastern
side.

The pumice flat over which we were now pursuing
our way, was frequently contracted to a narrow passage,
between vertical walls of the great environing crater,
on our left; and on our right, the frontal ridges
of lava streams, that have rolled down from the
central Peak in some distant age, violently surging on
thus far towards the circumference. These mere
edges of the streams are, in themselves, quite a range
of hills; and are extraordinarily rough, being a mere
congeries of great sharp-angled loose blocks of reddish stone, with little or no débris amongst them.

Here and there, at distant intervals, the grey
retama has taken root; but no other plant is seen or
makes any practical part of the vegetation. To travel
over such a surface with mules, would be impossible;
and happy it is therefore for travellers, that the lava
streams have not butted up close to the crater wall,
but have always left between it and them, some

small space, which has been afterwards silted up with
fine pumice. These flat strips of ground more or less
broad, are called "Canadas," with reference apparently to their valley position between the ridges on
either side. In winter the snow drifts deep in some
of them. Then they form a very treacherous, often
fatal, road; but unfortunately the only one, from the
northern side of this part of the island to the other.
Occasionally, magnificent blocks of rock, ten to twenty
feet cube, are standing isolated on the white sand;
and appear to have fallen from the crater cliffs above.
With a few such stones, we were continually thinking,
what a wind-proof residence could be made on the
top of Guajara.

Meanwhile the afternoon hours were rapidly advancing. The sun that had long been lost in the
region of the zenith, was beginning to shine into
our eyes from the west; our shadows were growing
particularly long; and the mules, who had known
no rest or stoppage since breakfast in the ravine,
were becoming very much fagged. The whole train by
this time had got into hopeless disintegration; there
were only five or six animals and men in a connected
line at any point; the different groups were generally

out of view and call of each other; we had even lost
sight of the dust-pillars of some of them, but were
almost choked in the clouds of our own.

A delicious story is now told us of a fountain, or
rather a water hole, which exists a little further on;
they even point out on the mountain-side a clump of
rocks, rather bigger than the average, but just as
brown and barren,—where they assure us that several
persons would be able to get a drink; and presently
these assertions are held to be corroborated, by the
appearance of two diminutive figures near the said
rocks. But we are rigorously deaf to any such be-
guilements, and declare that there shall be no stopping
until we reach the spring of Guajara; now however
not so very far distant.

The mural cliffs on our left, had been continually
becoming grander and more imposing; and at last
when we reached something of a bay or recess below,
and a depression at the top, we were informed that
this was the pass of Guajara. Here accordingly
began the work. We had now indeed to leave the
crater floor, on which we had been travelling so long,
and by clambering up the face, endeavour to reach
the summit of its lofty volcanic wall.

The slope was very steep, and the road or path

ascended only by acute zigzags. While climbing this ridge, the various groups of our cavalcade were again within eye-shot. The poor tired animals with their heavy loads did wonders; some of the Spanish boys too exhibited marvellous alacrity and endurance, despite the hot and arid air. At length the summit of the pass was gained, and winding round one of its heads towards the right, we came to a few hand-breadths of moistened ground. This was the long-expected spring, and here a half-hour's rest was awarded, before the last short, tough pull up to the top of Guajara.

The moisture in question evidently exudes from under a large face of rock, and has a tendency to soak on towards a magnificent ravine below us. This cleft begins at the head of the pass, and all at once plunges down with precipitous gorges, widening and deepening as it goes, and with many branches from the right and left. It runs towards the south, and is one of the many radiating fissures in the great crater wall, which are probably due to the violence of former up-heaval. Looking down into the blue depths, we see a large species of hawk sailing in mid air; and further down still, are some gaunt old pine trees projecting from a rocky shelf; but on our own level, vegetation is very scanty.

Presently, with a hissing whistle as they darted through the air, a flock of half-a-dozen pigeons just grazed our heads in their flight, and then curving upwards, flew away to the eastern mountains. At the sight, our interpreter, from a state of melancholy exhaustion, was instantly roused into ecstatic attention. " Oh," he said, after an interval of breathless suspense, " they have gone away quite now; they came here to drink, as this is the only water-place in all the country, but they were frightened at finding the spot pre-occupied." Poor little pigeons; for then he went on to detail what capital sport it was to lie concealed with a " musket" in an arbour of stones near this same water-hole, and shoot the said pigeons when they came to drink. Contrasting the case of these few miserable little birds here, with the uncounted variety of animals, graminivorous and carnivorous, that would be visitants at any similarly unique watering-place in South Africa, a vivid idea was gained of the respective zoologic riches of island and continent.

Time however being precious, neither art nor science could be indulged in then. The pigeons' visit was in itself a sign of evening, and the place where we then were, had almost lost the sun. The mules demanded our first attention, as their loads had to

be adjusted; and each animal was to be allowed a
short opportunity, of moistening its lips in the muddy
water, that slowly oozed into the holes which their
masters had industriously been digging, in the line of
humid earth.

The moment the specified time was exhausted, we
mounted again, and then began half-an-hour of good
hard work, as we pushed up the southern side of
Mount Guajara, amongst loose lava stones; and
amongst patches of dead-wood of *codeso*, often sorely
entangling the horses' legs.

Presently, in our winding ascent, we found ourselves
for a moment on the northern face of the mountain,
and the great Peak burst suddenly into view! The
sun was setting between it, and a high precipitous head
of Guajara, filling all the air and crater plain beneath,
with a flood of yellow light. The view was worth a
day's toil to gain; but could not, when gained, be indulged in for more than a moment. We struggled on
therefore into the shadow on the southern side again,
still upwards, and always upwards. Before long the
angle of ascent decreased, the ground flattened out, the
distances appeared on every side, and we found ourselves, a few minutes after sunset, on the summit of
Guajara.

An instant to look around, and then we find the air rapidly growing chill. Each Spaniard as he comes up with his mule, discharges its load anywhere amongst the rocks, and hastens away; striving, if haply he may reach the comfort of the spring below, before the shades of night descend. Ere many minutes have elapsed, each man has come and gone, disappearing ominously one after the other, below the mountain's edge.

We take another hasty survey. The light is faint in the western sky; but the moon rising in the east, encourages our little party, consisting of my intrepid wife and myself, Mr. Stephenson's two sailors, the Vice-consul's nephew, and two Spanish guides.

The situation does at first look somewhat desolate, but recalling geodetic experiences, I set the hands upon certain packages, and before a couple of hours are passed, two tents are erected, and a hot tea prepared, served out and done abundant justice to. The yacht carpenter, declaring that he is a regular washerwoman in the matter of tea, does not object to having his basin filled again and again. The air is fortunately nearly calm, and how clear; the moon is intensely bright, and yet the stars are also abundantly visible. By and by Jupiter rises, and appears to our northern eyes, as a brilliant luminary, that could never have been recognised before.

The day had been one of great toil, but this display of the nocturnal sky, showed that it had been all in the right direction; and here we were, within twenty-four days of leaving England, bivouacking at a height of nearly 9000 feet, on a mountain only 28° from the Equator.

PART II.

ON THE CRATER OF ELEVATION.

CHAPTER I.

SECURING THE STATION.

ON the morning of July the 15th, I rose by break
of day to examine the mountain top, on which
we had effected our lodgment in the dusk of the
previous evening. The prevailing colour of everything about, was light yellow; the soil, something like
a mixture of powdered clay and sand, plentifully
strewed with blocks of trachyte. These exhibited
generally some appearance of stratification; and the
ground about them was thinly, or rather distantly,
dotted with alternate bushes of *codeso* and *retama* ;
both of them very stunted, and oftener dead than
alive.

So far, there was nothing very different in the
physiognomy of the place, to what any ordinary hill-
top,—even the top of a very petty hill, of rounded
outline and sober unimposing aspect, as well as quiet
Neptunian formation,—might present. Yet not more
than ten paces towards the north, there commenced

the edge of a tremendous precipice, in its several ledges more than 1500 feet deep, running along all that side of the mountain, and forming in fact part of the internal wall of the great crater. The sudden discovery of this characteristic feature of a volcano, and in such close proximity to our tent, was calculated to make one tread the ground with more reverence. We felt inclined even to speculate on whether from the great Peak—the cone of eruption, rising up in dark and solemn grandeur in the centre of the pit before us—or from some of its subsidiary craters, we might not be made acquainted in the course of our stay, with a few other Plutonian features, and of an active character, for having presumed to pitch our tent on the very lip of this volcanic lion.

These however were eventualities so remote, that they quickly vanished before the practical require‑ments of the occasion; and these had reference mainly to meteorological causes. We had been most provi‑dentially befriended on our first night, by a calm atmosphere; but how long was that state of quiet to endure? In our present inexperience of the moun‑tain, we could not depend on anything; and well I knew, from seasons spent on ranges 5000 and 6000 fee high, that when the wind does blow there with a

will, it can—despite the greater rarity of the air—
produce far more destructive effects than down below;
so much does its velocity increase with height, if we
are thereby approaching the middle of an aerial
current. The wind was blowing violently enough at
the sea level, when we entered the Canarian Archipelago, a few days previously. And if by ascending
therefrom 9000 feet into the atmosphere, as we had
now done, we should be approaching the centre of
that aerial stream,—whose lower films, though retarded
by friction against the earth, yet lashed up the waves
into fury, and made the masts of the yacht bend like
whip-sticks,—why then we should certainly find Mount
Guajara, a rather untoward region, for carrying out
any very precise astronomical observations.

Reasoning from what I had already found to prevail in a similar latitude in the southern hemisphere,
and applying it to the northern latitude of Teneriffe,
the S.W. wind, which blows there in the winter, must
have a depth of far more than 9000 feet, and would
therefore produce a fearful climate on the mountain
at that season. But the N.E. wind, which prevails
during summer, ought, from the same precedents,
to have a depth of only 5000 or 6000 feet. Whence
it must result, that the head of Guajara, situated above

the strength and clouds of the lower, but not quite in those of the upper current, should then have placid weather, and enjoy its season of rest and quiet. This was theory; a theory that I had some confidence in, and which was somewhat confirmed at that moment by the calmness of the air about us, simultaneously with the immense quantity and rapid motions of the Trade-wind clouds, spread over the sea in all directions at a far lower level. Yet other persons, claiming equal right to an opinion, had threatened us with continued cloud, with driving sleet, and violent gales; some said from the north, some the west, and some the south.

This was the state of the question, when we opened our aerial campaign on the 15th of July; and, considering all our responsibilities, the first thing to be done, was, to secure the station. A mistake at the beginning, such as an unfixed tent being blown over the crater precipice by these threatened S.W. squalls, would have ruined our whole expedition; so I paced the top of the mountain over and over, arranging the best method of proceeding. The guy ropes of each tent had already been carefully fastened to large rocks as well as to pickets; and my wife's tent I had had constructed in Edinburgh on the following

plan, for promoting both cleanliness and safety.
Safety, in that the manner in which tents are generally blown away by a storm, and the inmates rendered most miserable and left most helpless,—being
by the wind getting in under the walls, and then
turning the thing inside out, like an occasional accident to an umbrella,—this source of weakness was
now avoided, by making a canvas floor; and forming it in the same piece with roof and walls. In
such case the wind, by getting underneath, cannot
enter the interior; and besides that, if all one's
heavy baggage is brought into this tent, it is so
securely ballasted thereby, as to be independent of
the usual pegs; and though it may be blown down,
that is the whole mischief. Cleanliness, at least in a
hot, dry, and dusty country,—because such a tent,
carrying its own floor, can at once be put down in
the sand, which then forms a soft understratum, but
cannot enter the interior in the usual way, and
begrime everything therein.

This was all well enough for a beginning; but when
the wind blows permanently with Alpine violence,
neither canvas nor rope can withstand it, without the
protection of stone walls,—a literally overwhelming

H

fact this, abundantly confirmed by Lieutenant Drummond, when trying his well-known light, on the
mountains of the Irish Trigonometrical Survey. Such
defences in solid stone, then, were to be erected here;
but as they could not, with our small party, be instan-
taneously raised, and as the opinions of informants
differed, with regard to the quarter from whence we
should first be invaded,—it became necessary to feel the
pulse of the atmosphere, and to try to judge from its
varied and multitudinous throbbings, in what mad
freak the winds were going to break out next, at the
end of the existing calm.

Anxiously therefore I watched the break of day, the
tints of colours on sky and sea, on clouds and distant
mountains; and noted with pen and pencil every
possible circumstance that could indicate, whether
this was an exceptional or a normal occasion. As the
dawn advanced, and the illumination became strong
in the east, the colours appeared somewhat weak
and sickly; dull, strontian yellow in the light, and
the shadows a poor gray. As sunrise drew nearer,
a blush of rose-pink surmounted the yellow, and
suddenly the upper limb of the solar orb itself, flashed
over the edge of the horizon, and across a plain
formed by upper surfaces of the Trade-wind cloud.

Though there was much in all this to excite one,
and though the picture was heightened in the foreground, by the Spanish guides in slouched hats and
long blanket cloaks, standing over a fire they had
made under a rock,—yet the impression was rather
unsatisfactory on the whole. There was an appearance of a dust, or dry haze spread through the
atmosphere, that marred the colours, as well as
lights and shades of distant objects. So presently,
when the sailors appeared, we set to our more mechanical work, and ranged all our packages in the
form of a square fence, enclosing, or rather standing
on, the very summit of the mountain.

Into this enclosure the sailors' tent was first
brought, for it was rather a frail and primitive affair,
standing much in need of protection. Its construction the previous evening did them infinite credit,
seeing that their materials were only a few of the
small beams prepared in Edinburgh for general purposes, and some pieces of old canvas in which some
of our boxes had been wrapped. These beams were,
on their second erection, so cleverly arranged by the
carpenter, in the form of an angular roof with braces;
that after they had been covered by the mate with
canvas, they were found quite strong enough to bear

the yacht hammocks being slung up on either side.
With these inseparable companions, swinging in ship-
shape style above their heads, some planks for a floor,
and the tool-chest for a table,—the sailors, though
raised 9000 feet above the level of their proper element, did not look by the second night, at all in the
state of castaways on a lee shore.

Hard work however was before them ; the walls,
the walls—every spare minute must be given by every
man on the mountain to this sacred labour, which
may or may not, presently prove the salvation of the
party. First, towards the S.W. corner, we began
our masonry; in the shape of a dyke about three
feet thick, and of as large blocks of trachyte as we
could move. The interpreter was fortunately skilled
in this dry-stone building, and the Spaniards were
consummate artificers; for it was precisely what they
had executed in the vineyards below. The sailors
also took zealously and kindly to the unsailor-like
work of carrying stones. Before long, all the good
blocks within and around the enclosure were used up,
and to save time in going backwards and forwards,
a barrow was needed. No sooner was the want apparent, than some of our generally serviceable beams
were sliced up by the handy carpenter, and put

together with a few screws into the desired shape.
The Spaniards excelled in the extraordinary dimensions of the stones which they occasionally brought up,
the sailors in the superior number of moderate-sized
ones, which they furnished, with the assistance of their
ingenious barrow.

At breakfast-time, I unpacked the simpler meteorological instruments. The state of the barometer was
lamentable; the one day of solar roasting which it
had experienced in coming up the mountain on the
mate's back, had blistered its varnish, and fearfully
twisted its mahogany legs. A little mercury had
escaped, but this loss was immaterial; and the instrument was soon properly suspended, and showing a
height of mercurial column of a little less than
twenty-two inches. In this there was not anything
very remarkable; for therefrom, either the hill was
not quite so high as expected, or the barometer was
standing particularly high, and was in so far a sign
of good weather. The temperature was pretty considerable, viz., 74° Fahrenheit, or not very different
from what had been experienced below. But in the
dryness of the air, what a change was there; the
depression of the dew-point from 10° in Orotava,
had increased up here to 50°. No wonder we felt

our lips cracking, our hair frizzling, our nails becoming brittle, and saw each other's faces scarlet; for the sun too was now shining down vehemently from a high altitude, where the dust haze no longer dimmed his rays.

A gentle air from the south at noon, and a breeze from the N.E. in the evening, completed our day; but even at 9 P.M. the atmospheric dryness was as great as ever.

July the 16th was much the same sort of day as the 15th, and we spent it in a very similar manner. It is described with this character in the astronomical journal :—

"Plain of cloud below, over the sea, seen very dense and compact; with an upper surface at probably 5000 feet of altitude. It does not come quite up to the island, which has attached to it a cloud stratum of its own, at an altitude of 3500 feet, extending towards, but not meeting, the sea stratum. Between them, the sea is visible; and appears curling in foam waves under a violent N.E. wind, while here we have only a gentle air.

"The horizontal strata of dust haze, at and above the level of our station, are very remarkable. They almost obscure the sun at setting, though the radia-

Photo-Stereograph 4.

TENT SCENE ON MOUNT GUAJARA, 8903 FEET HIGH.

p. 134.

Printed by A. J. McIntosh under the superintendence of James Glaisher Esqr F. R. S.
and published by Lovell Reeve.

tion in the middle of the day is excessive. No upper
clouds."

A sketch looking out of the tent door at midday,
shows the cone of the Peak, rearing aloft its mass of
thousands of lava streams and eruptions of pumice, its
terminal sugar-loaf cone, and crowning crater with
snow-white lips. Indications of the true colours
appear abundantly, but they are nearly overpowered
by the atmospheric blue of distance, and the dancing
sunlight that fills the whole region. This light glares
blindingly, from the Naples yellow of the ground immediately before us; witheringly, from the blue-white
of old dead bushes of *codeso*, reduced to the semblance
of white ashes, without the assistance of fire; and
transfuses through and through all the canvas of our
tent.

A sketch at sunset, shows the luminary entering
thick banks of dust haze, at the base of a perfectly
cloudless sky.   The Peak comes out then as a dark
purple mountain; and more conspicuously still, because
more towards the sun, appears the magnificent crater
of Chajorra, a huge caldron on the western slope of the
great Peak, that with a diameter of 4000 feet has
a height of 10,000.

All this is grand and soul-inspiring of course, but

there is an uneasy feeling difficult to define. What
mean those dense banks of dust haze, that completely
hide the horizon from us, and prevented us distinguishing the exact moment of sunset? Round about
the mountain beneath, we see the plateau of clouds;
they grow denser as the day declines, and at last
encompass us on every side. Piling one on the other,
they shut out the sea, the land, and all the lower
earth from our view; while they themselves in their
turn, as they stretch away into the distance, are lost
in the haze of the dust. Gazing at the scene, we
are forcibly reminded of the Mahometan cosmogony;
for here is the solid rocky mountain shadowed off
below and resting on the cloud, and the cloud again
resting on thick air or nothing at all. The air
though, is not that "still suffocating wind," that
formed the ultimate foundation of all things with
Mahomet; for as the darkness advances, squalls from
every quarter of the compass begin to assail us.
N., N.E., W., and S.W., are a few of the many
quarters from which mad puffs of wind came whistling out, shake the tent in a way to try all its fixings,
and then go howling off into the distance.

An extraordinary feature is now remarked in the
N.E., where underneath the horizon is seen a long

lurid streak of light. At least it seems underneath it;
for one can distinguish the dark banks of dust in the
atmosphere, gradually increasing in density with their
zenith distance, and becoming almost infinite on the
horizon; and it is below all this, that long line of
lurid glow. Are the dreadful eruptions of Lancerote,
which lies in that direction, beginning again, and reflecting themselves on the lower sky? The sailors
could give no guess at the meaning of the appearance;
but we prepared for anything. Carefully smothering
the fires with earth, to prevent accidents from stray
sparks, and looking once again to all the tent ropes,
we turned in at last with quiet and approving conscience.

The wild squalls long prevented sleep. But when it
did come it proved sound; for the sun was up when
we awoke, and what a morning was there. The air
was calm, and oh! so clear; every particle of the dust
haze was gone, and it seemed as if a veil, and a very
thick one too, had been suddenly removed from the
scene. The colours of even distant ranges of mountains were almost as vivid and pure, as on the ground
close by; and shadows of things, far or near, were
surprisingly dark and transparent. Now we could
examine the stratification of the opposite wall of the

great crater of elevation, the individual dykes of the Peak, or the formation of the distant island tops of the Canaries, Grand Canary, Palma, Gomera, and Hiero. The tops only in most cases, for the cloud sea, or strata of the Trade-wind cloud, concealed everything in the distance under a level of about 5000 feet.

At 7 A.M. on this day, a water-colour sketch represents the cloud sea to the east, as a dense white plain in the distance; where the sun shines on it, as white as snow. As secure a walking ground too does the surface of mist seem to afford, for no gaps appear in it, merely undulations of surface, caused by the close juxtaposition of numerous rollers of cumuloni cloud, stretching still from N.E. to S.W., exactly as we saw them when sailing below in the yacht.

On the E. and N.E., the clouds press close up to the land; but on the S., or under the lee of that part of the island, which rises to a greater height than their level, there is an interval of three or four miles. This enables us famously to see the edge, and judge of the thickness of this great platform of cloud. It is probably about 1000 feet, and produces under it, as we can also see, a region of darkness, that compared with the sunlight outside, looks *triste* indeed. Near the boundary of the stratum, a roller occasionally breaks off from the rest, and goes drifting away lengthways, and in a direction that indicates a strong Trade-wind there; though we, so much higher in the air, have a perfectly calm atmosphere. The same story also is told by the waves of the sea, which are seen by a telescope to be crested with foam; while a large steamer, ploughing its way heavily to the S.W., is preceded by its own smoke. It is accompanied likewise by so immense an amount of white frothy water, that we make sure it must be experiencing a double-reefed topsail breeze at least.

At 11 A.M. a view to the S.E. shows the purple summits of the many ranges forming the island of Grand Canary, sparkling above the level of the N.E. cloud; which is constant as ever in the distance; but declines to approach nearer the coast in this quarter, than four or five miles. The low country is somewhat dim and hazy; though what with the rocks and surf on the beach, the yawning mouths of many parasitic craters, and the winding angular lines of ravines, like so many huge cracks in the earth, there is no lack of detail anywhere. At the same time a ridge of the mountain, nearly two miles from us, and of about our own elevation, is throwing a shadow so dark, yet so

transparent, so black-looking amid the sun-lit landscape, and yet in reality so richly and deeply coloured,
—that a full idea can only be formed, by combining
what the telescope reveals of the powerful lights and
shadows in the moon, with the paintings of Turner
and of Rembrandt.

The day wears apace, and most luxuriously in so
pellucid an atmosphere, lit up by the rays of a vertical
sun, undiminished by any aerial impurity. Each
moment on a day of this sort is worth hours on any
other; we look at everything far and near, see it
as it were face to face, and gain a higher idea of
the glorious creation in which we live. As the sun
descends, the shadow of Guajara is thrown first on the
sea, then on the cloud plain; and then as the sun
actually sets, the shadow crosses the island of Grand
Canary, rises into the sky above it, and appears distinct and pronounced as a real mountain; and when
presently, almost behind the sharply defined summit
of the shadow, the full moon rises, and appears to be
the very cause of it,—the illusion is extraordinary
indeed.

Turn round quickly though, lest you lose the glories
of the west; see how vivid atmospheric colours can be,

when there are no impurities in the air. The chief part of the light is yellow, glorious cadmium, passing below into the richest tint of red-orange, above into lemon-yellow, and then powerful rose-pink, which by degrees fades into the deep blue sky above.

Underneath all this display is seen the Trade-wind cloud-plain, concealing all the distant sea, and forming the horizon on this as on every other side. Of a delicate blue grey, these clouds form so level a plain, that if only a footing could be obtained on the edge that approaches so close to Teneriffe, one could fancy it to offer an easy walk over to Palma, that rises on the verge of the glowing horizon.

While still admiring this scene, night arrives; one hour five minutes is the observed astronomical time of the duration of twilight, but that takes account of far fainter light than what practical persons would deem of any importance; and half that interval ought to see all wanderings about the mountain terminated. On this particular night a larger scope is allowed by the glories of a full moon, which soon abandoning the saffron colour with which it rose, mounts up in the sky as a disc of the purest white, and actually seems to stand out in front of the stars, and their background of indigo blue.

CHAPTER II.

SOUTH-WEST ALARM.

QUIETLY we had retired to rest on the night of
July 17th; and after a day so fully occupied,
slept soundly enough, little thinking of the morrow;
but the morrow came in due course, and proved quite
able to take care of itself, and establish its own claims
to attention. At an early hour, the shaking and
shivering of the tent, and the noise of wind increasing every moment, awoke us. We went out,
and lo! the direction of the gale was S.W.; the
threatened, and the promised, return current from the
Equator, had at last arrived.

If we must live in a wind, by all means let it be
the S.W., and not the N.E., that effete, unwholesome, used up, polar stream. As to the chemical
constitution and sanitary qualities of the two winds,
there could be no comparison between them; but
then, which was likely to do its spiriting most violently? We feared, after all, the south-west; be-

cause the heights were its proper province in these
latitudes.

Peering into the wind's eye, we could discover little
to guide us as to what was coming; the sky was
clear and blue, as usual; all the country below the
level of 5000 feet was covered in by the stratum of
N.E. cloud, that spread out over the sea as well, and
this was also its wont; again, all the country, craters,
and peaks, above 5000 feet of elevation, appeared as
dry, and as hard in outline as ever. The wind swept
down on us all the time, in a steady, unmitigated, unceasing blast, as if from a boundless reservoir; and was
now growing so strong that we could with difficulty
stand up against it. Towards the S.W., the horizon,
i.e., the cloud-horizon, was particularly smoky; and
sea-gulls, that had come thus far inland, retreating
before the gale, every now and then whirled up and
around; and then sped away, over the crater wall,
making apparently fruitless efforts to withstand the
violence of the squalls.

A cloud of dust was seen rising up from our
own hill, but presaged only the arrival of a small
party coming from Orotava with supplies. Gladly did
the muleteers unload their animals, and hasten down
to the sheltered ravine of the spring. One of them

waited for answers to the letters he had brought.
What said the first? It came from the most experienced person of the island in all that related to the
climate of its upper world. "Build your walls high
and strong," said the letter, "towards the S.W., or
your tents will be torn to ribbons." Pleasant comfort this, with the S.W. wind at that very moment
tugging and straining at every cord of our tent; and
that tent at the N.E. corner of the enclosure; or as
far as possible from the bit of wall already built, and
as close as possible to the brink of the crater precipice.

On reconnoitring once more outside, the Spaniard
was found under our small piece of completed wall,
wrapped up in his long cloak, hat doubled down, cigar
in mouth; and gazing with such an inimitable air
of lordly pity and sovereign contempt, at the frail-
looking materials of both roof and guy ropes of our
canvas abode, struggling and stretching as they were
in the unequal conflict with the growing gale,—that I
determined at once on the course to be pursued.
So answering the letters, and sending off the cynical
townsman, I called together our mountain party,—and
though the sun was high in the sky, as well as burning
hot, and my wife not desirous of having her domestic

arrangements for the day interfered with, yet for the sake of certainly being on the right side,—we removed her tent into immediate proximity of the highest part of our southern wall. Then unpacking some new coils of rope, we carried stays to projecting points in the solid rock of the mountain.

Next came consultations, how to bring this critical period of indefence to the most speedy termination, and our interpreter volunteering to ride down to Orotava, on the angular pack-saddle of one of the baggage mules,—being sure that it would be so dark by the time he arrived in inhabited places, that the women would not jeer him for riding with such equipment,—he was deputed to see how many men he could engage, to build off-hand all the enclosure required for our encampment.

With evening the wind died away. Then appeared the unusual summer sight for Guajara, of clouds; that is clouds on the sky above us, for there were the ordinary lower clouds over sea and land, as a matter of course. Long we gazed at the novel upper clouds sailing over our heads from the S.W. Of a delicate structure and *recherché* sort of character, they had none of the vulgarity of those great puffy clumsy rollers of the cumuloni cloud below. They were between what the

I

meteorologists call cirro-cumuli and cirrostrati, the
most picturesque of all clouds in the pleasing way;
equally eminent for their beauty of form, as for their
tender tints; and most appropriate in such semi-classic
scenes as the paintings of Watteau. Turner, would
hardly have been Turner, without these clouds to add
piquancy to the uniform blue of a mid-day sky, or to
reflect the glories of a setting sun.

These upper clouds sailed clear over the Peak, whose
height is 12,200 feet, and were probably not less than
15,000 in elevation. When they are seen by observers
below, there can be no *lower* clouds that day; and what
charming scenes of aerial cloud-land may there often
be aloft, when, as in Edinburgh, we have the N.E.
wind and its low clouds immediately over our heads;
spreading a murky and soul-depressing gloom over
everything; knocking out all the lights and shades of
nature, and replacing them by a cold, lifeless grey. On
such occasions the air is wet, yet it does not rain; the
icy wind meets you right in the face in every street, and
comes down on you with chill over the house-tops,
and from every quarter whence you don't expect it.
The pavement is slimy, and reflects the gas with an
unwholesome gleam. Every man you meet is closely
buttoned up, and is hurrying along with a dogged

sort of ill-will, against himself and everybody else; he looks neither to the right nor left, neither above nor below. Though he be painter, or though he be poet, yet he gazes not on anything; above there is only dulness, on the earth below only lifeless colours. Were he raised, however, at that instant but a few thousand feet in the air, the vivifying rays of the sun would be dancing about him; and if there were any clouds still higher, they would be these fairy-like cirri and cirro-cumuli, which may chequer a day and render it beautiful, but never make it dark.

As the flocks of these little clouds drove over the top of Guajara, some slight effect was produced by them on the dryness of the air; for the dew point depression was not quite so great as before. From 50° it had descended to 25°; a remarkably large amount still, and in which our hair continued to become daily more frizzly, our nails more brittle, and the wood of the packing-boxes went on steadily shrinking and cracking. The electricity of the air was frequently examined, but as on every previous day, it was found still small in quantity, and resinous in quality. Although we had had a touch of the S.W. current, the electricity of this region was still that of the Trade-wind, which is characterised always by extreme moderation and regularity.

On the morning of the 19th, finding that there were complaints of want of water, and no notion of any way of getting it, but by waiting for a man from below to bring it on a mule—I started off by myself, with a couple of tin-buckets to visit the fountain in the glen; *i.e.*, the wet place by the road-side which had proved of such service on the Monday. To walk on "foot-back," as the Dutch say, is the true method of becoming acquainted with a mountain; and I was soon rewarded by finding red, blue, and green lavas, small specimens of obsidian like artificial black glass; and, in the midst of a long slope of loose white pumice, a single lilac-coloured violet. Its underground stem was long, and the root distant from the place where the leaves and flowers appeared—so far had they been carried down by the descent of the soil during the time of their growth. Not being prepared for such a length of root-stalk, and hoping to meet with more specimens, I pulled, and broke it; but not another " *Viola Teydensis*" did we ever see again.

Directing my course by the bearings of cliffs on the opposite side of the ravine, I came presently on the water-hole of our ascent. Finding it rather muddy, I explored the neighbourhood, rambling over black lava and green-grey trachyte; until led to the origin of the trickle of water, in a cavern of small depth under a projecting ledge of white tufa.

On the floor of this hollow, was a little pool decorated with an array of floating spots of purple, as if the place was an abode of the smallest and most beautiful of all water-lilies. But what was my surprise on finding these gems of colour to be dead butterflies (*Polyommatus Webbianus*). An occasional specimen had been sporting in the neighbourhood of our station, but what made such quantities come to drown themselves here, it were difficult to say.

While engaged in filling the water-cans at a place where broken stones acted as a natural filter, there sounded the little tinkle, tinkle, that announces in Teneriffe the approach of goats. There they came trotting along; the more active ones jumping up all the rocks on their line of march; sometimes disputing with each other for the honour of standing on the very highest; and then running helter-skelter to be the first to arrive at another spot, suitable to their innocent ambition. There were milch goats among them; quickly, therefore, I emptied one of the water-cans,

and the herd-boy so readily understood my pantomime,
that he at once drove his flock up into the cavern, as
a *cul de sac;* and then catching one goat after the
other, poured glorious supplies of its rich milk into
the tin vessel.

I was rather surprised at the time, to see how readily
this youth acquiesced in my views; as well as how
liberally he supplied me: and not being able conscientiously to attribute it to my eloquence in Castilian,
had been inclined to give all credit to the British silver
which I offered him. We learnt afterwards, however,
at the station, that a neighbouring Spanish proprietor
—an admirer of science—Don Martin Rodriguez, had
given all his goat-herds instructions to be civil and
obliging to the strangers, if they should fall in with
them.

When three or four quarts had been drained into
the tin, the goats began to think that they had been
sufficiently taxed. First somewhat uneasy in their
far end of the hole, their impatience presently began
to manifest itself in attempts at escape; many a one
was caught by the leg, as it fancied itself bounding
into freedom; but at last one of them slipped out, and
then there was a regular *émeute;* and while the boy
and myself were stopping them on one side, they

rushed out on the other, like water through a sieve, until not one was left.

Still there was hardly as much milk as those on the station could utilise; so beating a large circle round about the goats, they were once more surrounded and driven into the cavern. By the time, however, that half-a-dozen more of them had been milked, the troop again grew riotous; and the bucket being now as full as was compatible with safe carriage up the mountain-side, the poor persecuted creatures were allowed to escape for good; and their herd-boy, with the money in his pocket, conducted them across the head of the ravine, to browse on the opposite hill.

With a bucket of milk in one hand and of water in the other, I followed the pathway to that wet place where our ascending mules had rested, and met a party of countrymen who were taking this way of crossing over from the south, to the north, side of the island. The shortest route this, but by no means an easy one; seeing that one amongst many other difficulties it included, was crossing a pass at an elevation of 8000 feet. Yet were these hardy Spanish peasants trudging along cheerfully, with large packages on their backs, to dispose of in the port of Orotava.

Being curious to see what sort of produce could

repay this toilsome mode of carriage, I applied to one
of the men who was resting himself. Most good-
humouredly and smilingly he took me to his pack,
lying on a rock hard by, and consisting in a flat painted
box, about two feet square and six inches thick.

On it were a number of hard green pears, his own
food, and of which· he freely offered me. But I thank-
fully declined them, perhaps not without some
epicurean dream of there being fine luscious grapes
within. So he proceeded to unbuckle the broad black
strap; and then, throwing open the lid of the box,
displayed its interior divided into a dozen partitions,
each filled with little grey particles like the ashes of
a cigar! These were the cochineal insects, as picked
off their cactus plants, and prepared for the European
dye-market. An admirable article of produce for
mountaineers without roads, for it is light, compact,
and high-priced to a degree;—while the market
has never yet been overstocked, nor is it ever
likely to be, so long as young damsels have a taste
for pink ribbons;—but then what a dish to set before
a thirsty traveller under a hot sun.

The unexpected supplies of pure water and rich
milk· which flowed into our mountain camp that

morning, infused new energy into the wall-build-
ing party.   My wife found her English prejudice
against goats'-milk rapidly declining, when she had
it thus fresh and unsophisticated; and the men felt
incited also to go out after their day's work should be
over, and see what they could forage for themselves.
The first attempt, however, was not well directed,
and I even felt obliged to express formal disapproval
of it; and, although assured that it was the practice
of the mountain, to order that it should never be
tried again by any member of our expedition.

The practice must doubtless be pleasant enough
for travellers, but particularly otherwise for Canarian
farmers; for it consists in catching and slaughtering
one of their goats, that run half wild and half tame
over the upper part of the mountain.   They are there
during the whole of the summer, with little attention
from their masters; for they are all marked, and are
sure, on the setting in of winter, to come back kindly
enough to the warm farmsteads below; and there are
no animals of prey to thin their numbers.   Happy
African isle, without the jackals, hyenas, leopards, and
lions of the continent to which it belongs.

Picturesque it is sufficiently, to see two bronzed
mountaineers, by the light of a flaring fire of brushwood,

in a crevice of the crater wall, flaying and cutting up, grilling and roasting wholesale; and afterwards to find a party under a gipsy tent feeding *ad libitum;* while from a projecting spar outside, there is swinging in the dry wind, and moonlit air, a dark haunch of goat. But from all this poetical brigandage, our astronomical party was entirely to abstain: and by moral example, humbly to aid and abet the general cause of law and social order in these highly elevated regions.

Next day being Sunday, and after six very exciting days of hard work, each man being left to his own ideas of passing the seventh—a more innocent excursion than kid-hunting was got up by some of them, in the shape of a walking expedition to the village of Chasna; about 4000 feet up the southern ridge of the mountain, and 5000 feet below our station. To this place started off one of the sailors and Manuel "the Marquis;" the other sailor made it literally a day of rest, and in the hot still air of the afternoon, his stertorean breathing resounded over the place.

About nightfall the Chasna expedition returned; the Englishman thought it a very poor place, just half-a-dozen ordinary houses, he said, stuck down by the road-side, and the doors of all of them closed.

But he launched forth, in untutored admiration of
the natural scenery he had been through : the deep
ravines, the yawning mouths of craters, and their
colossal heaps of coal-black cinders, as fresh and
natural, as if they had been thrown out only yesterday, from some burning fiery furnace.

CHAPTER III.

THE method of "term," or stated days, whereon meteorologists should observe throughout all the twenty-four hours, was an improvement introduced by Sir John Herschel, when studying the climate of South Africa. So admirably is it adapted to bring to light phenomena of short period, that I was happy when the morning of July 21st arriving, enabled me, with a semblance at least of utility, to begin the hourly observations of that particular term-day. At six o'clock, accordingly, the first entry of the series was made.

Our meteorological observatory was now pretty satisfactory as to arrangement, and was practically efficient in guarding against local disturbances. These were mainly two; first, the terrific radiation of the sun by day, with something of the same sort, but with an opposite sign, from the sky by night; and, secondly, the wind, which, if allowed its own sweet will, might at any moment blow all our instruments into the crater. To protect from these, a portion of a room had been erected, with walls to the S.E. and W., and with a bit of roof formed of planks and canvas, covered in with brushwood and weighted with rock. This roof was to keep out the almighty prying of a vertical mountain sun, whose rays did indeed try the nature of everything they had access to. The walls were of stone, and pretty nearly four feet thick, so that there was no fear of direct solar radiation getting through them. Yet as they were at the same time built loosely with angular masses, they were quite porous to the wind; which kept nicely circulating in and out amongst barometers and thermometers, so as to prevent the formation of any local atmosphere, and enable us to obtain the true temperature of shade on the mountain.

Seven o'clock, eight o'clock, and so on through the day, saw us repeating our observations of the several instruments, as well as unpacking and erecting the Sheepshanks Equatorial. This was the smaller of the two telescopes we had brought to Teneriffe; and with its stand, packing into four heavy boxes, was no mean load for the mules. I had decided at last on trying this instrument at Guajara, in place of

attempting the doubtful success, and expending all
our means in at once bringing up the larger Pattinson
Equatorial. Amongst other reasons, there was already
looming in the distance, the ultimate necessity of
reaching a higher station than this, in order to be
above " dust haze." We must first ascertain how much
we can do here, I reasoned, with moderate means; and
will afterwards use our utmost exertion, and spend our
last shilling, in getting the more powerful, but impossible telescope,—as our Orotava friends call it,—up to
the greatest available height along the slopes of the
Peak.

Could any other conclusion well be come to, when
the abnormal impurities of the atmosphere at our
level, were almost daily described in the meteorological
journal in such words as these:—

" Air still very hazy on horizon; haze when under
the sun, or in a position to reflect light, appears
whiter and brighter than the pure sky, and when
seen in front of, or by transmitted, light appears
darker. At sunset some small cirro-cumuli, seen
through a haze stratum in the west, itself dark as
smoke, were vividly bright."

Again: " Air thick with dust haze; distant mountains consequently indistinct; except where, as with

Chajorra and the Peak, they rise above 9000 feet
of altitude; anything above that height being seen
sharp, black, and distinct. Night after night, when
all lower things are oppressed with dust haze, the
summit of the Peak appears to shoot up through it
into the clear blue vault of heaven."

Because so much is said of this "dust haze" on the
mountain, let no one fancy that it is an impurity,
from which an astronomer would have been safe, had
he remained below, for it spreads through the air
at one definite level for hundreds of miles, far and
wide, and quite independently of the Peak. An
observer therefore at the sea level, would equally
have the untransparent medium between his eye and
the stars in the zenith, as would a man at a higher
elevation. Towards the horizon, however, there will
be some difference; for if the hill station be nearly at
the same height as the layer of dust, its apparent
density will increase, from perspective, so rapidly
towards 90° of zenith distance, that it will form in
that part a very visible band; while at the plain
station, there is no such rapid increase of perspective,
consequently no very great variation of visible effect.
In low countries there is accordingly many a "pale"
blue sky seen by day, and there are unaccountable

absences of the zodiacal light or milky way noticed
at night, which are owing precisely to the advent
of a thicker stratum than usual of this said dust haze
in the upper air, without the inhabitants below sus-
pecting the reason. On Guajara, however, where we
were nearly on the level of these dust strata, we could
always detect their presence, and judge of their
density over head, by referring to appearances in
the neighbourhood of the horizon.

Seeing then so clearly these beds of dust, spread
abroad in the air, we were inclined to talk of them
even over-abundantly. "That which is *unexpected*,"
says Arago, "always becomes the lion in scientific
researches." So having ascended this mountain for
the purpose of getting above clouds, and as much
of the atmosphere as possible; viz., its nitrogen,
oxygen, carbonic acid, and watery vapour, we found—
on beginning operations at 8903 feet of altitude—
there was another difficulty in our way besides
aerial gases, in the shape of these strata of " dust
haze," or of finely divided particles of solid matter.
Straightway, then, these became our chief source of
anxiety to understand, and of ambition to overcome.

While engaged with the sailors in unscrewing the

telescope boxes, I looked up, and lo! a little bird
of the fringilla species,—a very little bird as compared with any that a working man would generally
care for,—came and perched on a stick that had been
fixed in the ground close by, to show the site of an
intended angle of the wall.   A tame innocent bird it
was, not unlike a common linnet; it hopped about as
far as the confined area of the stick top would permit,
now to the right, now to the left; then stooping down,
wiped both sides of its bill cleverly on the edge of the
wood; looked up at us archly and merrily again; and
once more began hopping backwards and forwards as
at first, in azimuth, if I may be allowed so to say.
In fact it performed its part so prettily, that with my
head full of expected canaries on this, the highest
island of the Canary group,—I asked our interpreter
by what name he called this exemplary little bird of
the brown plumage. He looked at it, and started, quite
sportsman-struck, as if he had seen a stately antelope.
Without answering my question, he called Manuel's attention to the diminutive fowl, not five feet off, and he
looked also—then they both had a long and animated
discussion; at the conclusion of which the interpreter
turned round, and assured me, with infinite *empresse-
ment*, that " the bird was remarkably good eating;

K

and that if he had only his musket with him, he would shoot the creature then and there.''

This did not say much for the amount of game, or of shootable things in the island generally; and living creatures of any sort were not very plentiful on our lofty abode. But we had already noticed a hawk, several swallows, some long-legged spiders, two species of large and beautiful flies, besides the ordinary sort, a small blue butterfly, lizards and grasshoppers. Moreover, those astute foragers, the crows, had found that something was going on which would afford them pickings; so they began to arrive from distant parts, and examined us narrowly from safe rocks in the neighbourhood, while organizing their plans of procedure.

A photograph taken late in the afternoon, shows the Equatorial mounted, and approximately in position. Its stand, in the shape of a hollow pier of wood, filled with stones to make it heavy, gives promise of resisting the wind. The two sailors are seated about amongst the packing boxes, looking very tired. One of the guy ropes of the tent crosses the foreground, and in the distance is the magnificent Peak of Teyde, raising its sugar-loaf cone high into the sky. At the foot of the cone, or at an elevation

Photo - Stereograph 5.

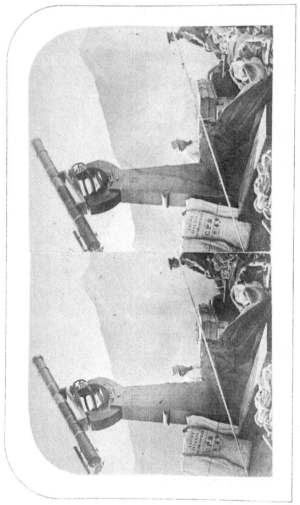

SHEEPSHANKS TELESCOPE FIRST ERECTED ON MOUNT GUAJARA, THE PEAK
OF TENERIFFE IN THE DISTANCE.

p.161.

Printed by A. J. Melhuish under the superintendence of James Glaisher, Esq.F.R.S
and published by Lovell Reeve.

of 11,700 feet, there is still a patch of last winter's snow; and below that begin on every side the streams of lava and pumice, various in colour, but subdued by distance into good keeping for the background of a picture. (*See Photo-stereograph, No. 5.*)

"Delighted and surprised," says our astronomical journal for eight o'clock on this evening, " with the marvellously fine definition of all the stars seen in the telescope; they all have such perfect discs and rings with a magnifying power of 150, a thing I have never witnessed once in Edinburgh, with this instrument."

Then seeing the constellation of the Great Bear going down behind the Peak, and having a sad hankering after a station upon its slopes,—yet fearful of the propriety of the place, from what a learned traveller has written of the vagaries he saw, while there, in the motions of heavenly bodies,—I watched immersions of different stars, with full power of the telescope; trying to detect any symptoms of extraordinary refraction, such as might arise from emanation of hot vapours at the occulting ground.

This is the result as entered in the journal :—"At 19h. 55m. Sid. time, saw γ Ursæ Majoris occult behind the Peak of Teneriffe, at about the level of

Alta Vista; the star showed a good shaped disc with well-formed rings, through the whole of the time it was watched; though prismatically coloured; red above and blue below. The occultation was sharp and precise.

"At 20h. 18m. 25s., δ Ursæ Majoris went down behind the very crater of the Peak, and similarly with γ.

"At 20h. 38m. 2s., α Ursæ Majoris went down coloured, but with good-shaped disc and rings, behind the sugar-loaf or ash-cone of the Peak; but the immersion was not perfectly instantaneous."

And this conclusion was appended to the above:—

"Considering that these good immersions took place at altitudes so low as from 0 to 9 degrees, there appears no reason to fear any excessive disturbance of vision by the hot vapours of the Peak, even were the telescope mounted on its flanks. While if thereby a greater height of station could be obtained, immense advantage would result in getting above the dust-haze medium; which seems, to one who is above the clouds, to be the chief remaining obstacle to a clear view of the heavenly bodies."

Meanwhile the night wore on apace; at every hour the meteorological instruments were noted. The temperature, which had risen to 74° at 2 P.M., went

down to 61° at 9 P.M., at midnight was 57°, and
shortly before sunrise was 55°. The dryness of the
atmosphere, as shown in the depression of the dew
point, followed more clearly the sun's motion; for it
reached its maximum, or 47°, at noon; and nearly its
minimum, or 32°, at midnight.

Daylight showed the cloud-sea as unusually extensive; and so closely thatching in all the ocean and
lower lands of Teneriffe, that not a particle of these
was anywhere to be seen. To all the natives below, it
must have been a regular, unmitigated, gloomy and
cloudy morning; though probably not worse than
that species of daybreak, which a Scotchman, habi-
tuated to—and to worse—calls "a fine grey morning."

To us on the mountain, there were no "upper
clouds," and the morning was overpoweringly bright;
yet the air was "dusty." The duration of twilight
was one-fourth shorter than on the previous evening;
and when the sun did rise, he was seen very pale, in
the midst of a dry haze bank, whose upper edge was
still 1° above him.

Having thus witnessed the day of the 22nd begin
its course, and also seen our men, now strengthened
by a party of six Spaniards from Chasna, commence

their work at the walls, I retired into the tent to sleep until breakfast time.

The Spaniards understood their part so well, and laboured at it so energetically under the direction of our interpreter, that before the close of the 23rd, we were enabled, with safety, to replace my wife's tent in its original N.E. corner of the enclosure. From thence the door gave us a charming view of the Peak of Teyde, its rugged dykes, and its lava streams of various colours. At the back, and on either side of the tent, ran dwarf stone walls, to keep off the wind; the narrow intervening spaces serving as larder and pantry. (*See Photo-stereograph, No.* 4.)

In front of the door, thanks to the laminated, if not stratified, nature of the trachytic lava, we mounted a large slab of it, to serve as a table; and a little further off, constructed a fire-place. Eventually we had several fire-places in different directions, to be used accordingly as the wind blew; for in so violently dry an atmosphere,—very often with 50° of dew-point depression,—and living under canvas in a breezy locality, no amount of caution against stray sparks could be superfluous.

The supply of fire-wood was inimitable. The ground

was dotted over with alternate bushes of *codeso* and *retama;* and each plant vied with the other for burning brightly, even when in the green state; so what would they not do, when dead,—as every other one was,—and so dry, so astonishingly dry, as they invariably became. Then there was such a nice distinction of fuel properties between these two shrubs; the *codeso* with its stringy fibrous bark, and the innumerable needle-like processes of its twigs, being so splendid for beginning a fire; and the *retama*, with its thick, solid, smooth branches, stems, and roots, so excellent a material wherewithal to keep up the blaze.

I have made bigger and hotter fires with *doornbooms*, and trunks of other trees drifted down with the flood of a South African river; and in the karroos there, have made larger and brighter flames with some of the dried euphorbiaceous plants,—but for good useful, semi-domestic fires, to cook the pot by day, and to make the night cheerful,—fires not so small as to keep one constantly attending them lest they go out, and not so large as to make one fearful of possible consequences,—commend me to the dual bushes of Guajara, *Adenocarpus Frankenoides,* and *Cytisus Nubigenus.*

Of the latter, only one or two specimens were in flower, its season being past : but we saw quite enough to understand how, when all the tribe was in blossom, the townspeople must find it well worth their while to pack their hives of bees, as they do, on mules, and bring them up to these higher regions; to gather honey from the innumerable cream-coloured and broom-like flowers of this famous plant. The *codeso* was still in full vigour of florescence, and its bright yellow blossoms and aromatic leaves, lent point and interest to the rocks as one walked along.

Other fires had to be considered in our domestic arrangements, besides those of artificial combustion, for the vertical sun shone down so fiercely on our tent by day, that had it not been constructed with a great part of the upper walls to let down on each side, and so permit the wind to blow through and through,—it must have become an oven of insupportable heat ; and my wife therein, would not have been able to lend me the important assistance she did, in writing and many other occupations.

The sailors' tent did not easily admit of such an arrangement; but they rigged up a sort of arbour with spars and canvas just outside their door. And at night, they showed the best disposition to take

things pleasantly, far more important than any material means, to enable them to bear the heat that then
arose from below. Their circumstances must have
been rather trying, for this was the sort of dormitory
arrangement. The seamen climbed up into their hammocks, very nautically slung in the upper part of the
tent; then the six Spaniards crowded in below, and
sent up fumes of garlic, that were quite astounding to
the unsophisticated senses of Britishers.

However, they all made the best of it, and got on
famously. The carpenter, when he saw the great
wall surrounding our whole station, and afterwards
smaller walls intersecting its interior, and affording
special protection to each tent and retreat,—began
to say that they were being made so comfortable,
that they would not like to leave the station. The
chief Spaniard, too, was so proud of his own share
in the work, that he talked of erecting a cross on
the highest corner-stone, and of carving his name
thereon.

An opportunity now occurred which enabled the
whole party to revel in good things, for a pedlar arrived, bringing along with him a mule-load of fruits;
delicious purple figs, allowed to hang on the tree until
fully ripe,—a justice that strangely is not often ac-

corded to them,—sugary, luscious, and of the finest
flavour. Then there were the most exquisite plums,
purple and yellow, of fine oval forms and growing in
such charming groups and bunches, that a cluster of
them,—with a few green leaves appended, as they
generally were,—would have been a matchless prize
for a fruit painter. The plums came from Chasna; the
figs from Grenadilla, a village lower down and in a
warmer zone. Then from somewhere between, came
immense basketfuls of the fruit of the prickly pear.
The Spaniards devoured them wholesale, that is, such
part as can be devoured, viz., the small quantity of
juicy pulp in which the seeds are enclosed. A pleasant flavour truly had this portion, tempered too
with such a degree of gentle acidity, as to make it
most efficacious in quenching thirst; but the British
sailor looked at it with excessive disgust, and denounced it as so much sheer nonsense in the shape of
eating.

Pears came from the same quarter, but had generally to be boiled up with sugar, they were so hard
and unripe; even the crows had much difficulty in
selecting specimens that pleased them. At least,
they got hold of our fruit-basket one morning before
we were up; and drove their beaks into every pear

there, just as a grocer would drive his long steel taster into a Gloucestershire cheese.

Fruits, notwithstanding, were the most easily kept of all our stores; bread perhaps the most difficult, for in half a day it became so extraordinarily dry, and desperately hard, as to need teeth of iron to make any impression upon it. And the meat? why the difficulty was not so much the keeping it when once obtained, as obtaining any to keep. The frugal Spanish peasants rarely touch meat themselves, and do not rear it for others. But one morning our interpreter came to us radiant with joy, for he had found a proprietor in Chasna, who had, he said, a " ram "; and this ram he was willing to kill, if we would take half of it; the other half he hoped to dispose of amongst the Spanish invalids and fashionables who were then flocking to Chasna, for the sake of its coolness and medicinal springs. A half of the alleged ram came in due course of time, but proved to be little larger than half a hare ; while in flavour and in fact, it was happily very excellent lamb.

My wife suddenly reported one evening, that smoke was proceeding out of part of the mountain ! Every one turned out to look, and with rather serious coun-

tenances.   There, in the basin of the great crater,
almost vertically under our station, lay a grotesque
collection of pinnacles of volcanic rock; and from some
of these, every now and then, fits and flashes and
wreaths of smoke arose.   (*See Photo-stereograph, No.* 6.)
There was not so much then as there had been, we
were told; and we had to look long after each appearance, before we could verify it by another.   We were
gazing down on the scene, at almost as great an angle
as if it had been in a map, while the forms of the
rocks were so picturesque, and highly coloured, that
they might have been in fairy land.   It was soon
tolerably plain, from the thinness of the smoke, that
if volcanic, there was nothing very violent going on;
and presently a theory was broached, that goat-herds
had been setting the *retamas* on fire.   This idea at
once relieving every mind, a general rolling and
throwing of stones down the precipice began.

One of the men, Manuel, or the " Marquis," especially distinguished himself.   The mildest and best
dispositioned of men, he had all along performed his
part most admirably at the walls.   The earliest to
begin, the last to leave off work, and carrying always
the heaviest stones.   Now, after all the labour of the
day was over, and a little play beginning in the dusk

of the evening, he commenced exerting himself like several men all in one, to roll down a monster block. Then to see the poor fellow all this time, with his hands bound up in bundles of rags; for what with a preternatural roughness in all the igneous rocks about, and such astonishing dryness in the air,—every worker on the mountain, native as well as foreigner, had got his nails broken down to the quick, and split up besides; while his hands were covered with lacerations, to say nothing of blistered face, and split lips. Poor Manuel's hands were in a worse state than any one else's, yet he brought up the big block successfully from a considerable distance, and finally launched it over the edge of the cliff. After five seconds we heard a tremendous crash, then a series of others, as if the block had broken into many pieces, which were still bounding down precipitous ledges.

There was yet something peculiar in the fitful way in which smoke occasionally curled up from the "lunar rocks;" and they looked so small, and so close by, that the carpenter was determined to be off, and examine the thing at the place. So away he went, to descend by a side ravine, and not to come back until he had actually felt, with his hands, whether the ground was hot, and giving birth to volcanic smoke.

Some time after dark he returned—completely crest-fallen; every essay that he had made, had only resulted in discovering the brink of some yawning chasm; and all his walking, and all his scrambling had not sensibly brought him any nearer, to the really far-off line, of " Los Corales."

CHAPTER IV.

THE GREAT CRATER.

THE "Lunar Rocks" had been a subject of high admiration, and intense puzzle, from the first day of our tenanting Guajara. We had complete command over them as to view; for, from the top of our cliff, or volcanic wall, we gazed almost vertically down upon them as they lay, or rather rose, and shot up, on the floor of the gigantic crater. Was it as a Gordian method of solving the difficulty of their origin, that Von Buch left them entirely out of his large map; otherwise a most excellent one, specially called a *carte physique*, and engraved so laboriously in the line manner, as to leave no part of the paper unburdened with careful conventional shading. Suffice it that they are not there; yet were in earlier Spanish charts; and are also in that more recent map of MM. Barker-Webb and Berthelot, which the friends of the great German geologist have so unhesitatingly condemned in everything.

Look down at any time of the day from Guajara, and these "Lunar Rocks" riveted your attention. In the middle of the day they gleamed again with bright greens, reds, blues, purples, and whites, as well as with yellows and browns. The greens, and likewise all those other colours being due, to the nature of the rock. Not to vegetation certainly; for of that there was practically nothing, even in the plain about, save the barely visible, far between dottings of globular *retama* bushes; the living, dark grey; the dead (*see Photo-stereograph, No.* 6), cinereous white.

To study these rock-formed enigmas from their commencement, we must come to the very edge of the precipice; and as the air about us rejoices in little puffs and eddies of wind, it will be as well to lie down at full length, allowing only one's head to project over the brim. Then looking vertically downward, we see on the floor of the crater below, a long ridge of rubbish, coming out as it were from under the cliff; and at intervals along the line, which trends generally towards the great central Peak, there shoot up some remarkable masses of rock. Rocks simply they can hardly be called, because they are so large; and hills would be more inappropriate, because they are such

Photo.—Stereograph 6.

CLIFF AND FLOOR OF THE GREAT CRATER. 8 MILES IN DIAMETER. AND 7000 FEET
ABOVE THE SEA, UNDER MOUNT GUAJARA.

p.144.

Printed by A. J. Mildmush under the superintendence of Thomas Gravior, Esq.", F. E. S.
and published by Lovell Reeve.

strangely steep vertical forms, of nothing but sheer hard stone. Following this ridge along with the eye, its *debris* portion gradually sinks and is lost; but its rock-masses continue to shoot up more strangely, and as it were more unreasonably; for they start forth at once from the ground without previous notice, and with perfectly vertical sides.

There is a square tower amongst them, more than twice as high as it is broad, which might almost pass for a specimen of masonry, seeing that it is composed of horizontal strata; but then one side of the tower is somewhat convex, and the other concave; and the breadth rather enlarges, than diminishes, towards the top. On the nearer side of this pillar, is a curious triple structure; while on the further, is a gorgeous collection of spires, minarets, ridges, and sharp-pointed peaks, glowing in varied hues; their bands of colour, as well as the weathering, all indicating a horizontal stratification.

On the flat near the square tower, we see something like a tea-party, rather gigantic, but forming a friendly, sociable group, and apparently discussing the scene. On closer scrutiny, the guests turn out to be needle-shaped rocks, sitting up as abruptly, as so many nine-pins. In the evening and morning

L

the long shadows which are thrown athwart the plain,
—by the members of our "tea-party," as well as the
cathedral collection of spires and towers,—are such as
one can only compare, with those which a telescope
reveals in the moon, at certain periods of its solar
illumination.

Coming fresh from the glacier and iceberg discus-
sions of Scottish geology, these "Lunar Rocks" had
first reminded us of the "grand mulets" on Mont
Blanc, with their glacier removed. For on the eastern
side, was something amazingly like a series of old terminal moraines, one behind the other; showing apparently how some ancient ice-stream, all of the olden
time, gradually waning before a secular increase of
terrestrial temperature, had finally vanished from
earth, and departed into air; leaving behind it, all
the way up the mountain slope, long winding lateral
moraines, which had once measured the breadth of its
frozen river.

Learning, however, to throw old associations on
one side, and attend only to absolute facts, before us at
the instant—we traced every apparent glacier-marking
so completely up to the various craters,—to the huge,
and it may be called the still active one of Chajorra,
as well as the equally large and unextinct, but yet

filled up one of Rambleta,—that there was no doubt
left in our minds, as to the long curving ridges of
red stones, in front of the "Lunar Rocks," being
in reality, the termination of a stream of lava.

When travelling through the Great Crater, during
our ascent of the mountain, the stupendous scale of
everything prevented our making out general features.
We were cognizant then, only of long rugged hills of
horrid stones; *i. e.* broken, angular blocks, sticking
and bristling up in all directions, and could make out
little more. But now, from our commanding height,
and bird's-eye view position, all those asperities of
surface were smoothed down, and we saw the con-
secutive wrinklings of the lava streams, as plainly, in
fact, as if they had been painted in a picture.

Day after day we gazed at, sketched, and discussed
these various outpourings which had flowed down
from the central Peak, deluging the plain of the
great crater; and insensibly we glided into a generalization, which further experience has fully confirmed. It may be stated thus :—

The earliest lava streams are of a yellow tint, the
succeeding ones red, a rich Indian red, and the last
are blue-black. The *yellow* appear to have been the

most abundant, as well as most fluid; for they cover
the largest spaces, have flowed over nearly level tracts,
and their ridges imitate the forms of watery waves.
In one of our photographs of the south-eastern
corner of this broad crater, the confines of a flood
of yellow lava from the Peak, may be seen rushing
up the curving beach in surf-like waves, as with the
sea on the coasts below.

The *red* streams, again, are evidently much smaller in
extent than the yellow, and have never run or spread
very far. Their terminal markings are more like the
wrinkles of a glacier, than the waves of water; and,
besides these transverse features, there are beginnings
of a longitudinal arrangement; in some cases, as
mentioned above, looking like the lateral moraines of
an ice stream. In others, they give one the idea
of nothing so much, as the ruts of chariot-wheels
of Grecian demigods, driven with celestial power
through the bewildered plain of loose red stones.

The *black* streams, are decidedly the scantiest of
all; they have never moved, except when the slope
was very notable; and with them, the longitudinal
arrangement, which had just begun to appear in
the red, predominates; all the black streams, being
nothing but a series of long ridges or embankments.

They have not the form of any sort of fluid stream, watery or viscous, but rather of a quantity of finely-comminuted solids, as sand; their sides, and even their ends, being sloped so uniformly at a constant angle, that they look here and there, amazingly like embankments formed by railway navvies.

I do not propose here, to enter into minutiæ of the absolute manner of movement of a lava stream, and the oft-discussed influences of viscosity and crystallization in modifying its manner of flowing; but only to point out differences of shape, on the large scale, actually subsisting amongst different streams. These shapes, being undoubtedly an expression of the particular mechanical forces once exerted in each case—must be replete with instruction, if rightly interpreted. Their study constitutes, indeed, a sort of colossal or telescopic mineralogy, which assumed in my eyes quite an aspect of professional importance, as presenting the only means, by which we can legitimately compare the surface of the moon, with that of the earth.

The relative ages of streams, alluded to in the enumeration already given, we ascertained by their position. The colour was an accident, or at least was superficial; but the differences of form were something of far greater importance, and when taken in conjunction with other features,—also capable of accurate measurement, as, relative extent, quantity, and angular slope of the bed,—indicated, besides their age, the gradation of heat in the different classes of streams, and showed, at least with this volcano, that a secular progress had accompanied its periodical movements.

Why certain leading geologists will so perseveringly refuse to the earth any secular change, over-riding and permeating its periodical movements,—when, from the range of physical astronomy to the boiling of a tea-kettle, we see that any effect of long period, is always mechanically bound up with others of short period,—it were difficult to say. But this Peak of Teneriffe, that is the central cone, or crater of eruption—in the midst of the vast crater of elevation, which will afterwards form an extension of our view,—shows indubitably, secular and periodical qualities co-existent.

Examination of the streams which have been sent forth within the memory of man, would tell little of secular progression; for they only break out about once a century, the last eruption having occurred in 1798, and the previous one in 1703; and it is necessary, therefore, to question, as we have done, pre-historic phenomena. This method at once enables us to take up a far more powerful position; for what a mighty period must have elapsed, to include alone the thousands of black streams, which now seam the cone on every side; and finding many of these, that are quite beyond the memory of man, untouched by any oxidating influence—we are inclined to wonder what myriads of ages must have further elapsed, to produce the deep red and yellow decompositions, so conspicuous on the surface of still earlier streams. For their blocks are, in the interior, black; and in chemical composition, very similar to the substances last ejected.

We have here at all events, whatever the absolute dates, results from an immense duration of periodical effects, spread before our eyes; and cannot resist the conclusion which their forms set forth, viz., that a most marked secular change has been wrought out, and that its signs are visible still. Hence, we may state as the general law, holding good on the whole,— not, of course, in every individual instance, any more than that the tide-wave advances uniformly, and without little undulations on its surface,—that the earlier streams were the most copious and most fluid;

nay, we may pretty safely add, the hottest also, and that there has been a continual decrease in size and heat ever since. The Peak of Teneriffe has, in fact, been steadily burning out for ages; and is, happily for mankind, no longer in its youthful energy, nor in its primeval vigour of destructive power.

In taking pictures of the several volcanic pheno- mena, our camera and photographic tent had been blown over more than once. Certain mahogany grooves being thus broken, for the replacing of which our spare deal-wood was not sufficiently strong, the yacht carpenter contrived to supply them by ingeni- ously cutting up a tent-peg; while he manufactured a new hinge, out of a bit of iron hoop and a nail. Other unlooked-for accidents would often occur, amongst the most frequent of which, was the opening of cracks in camera-box, or plate-boards, in consequence of the desert-like dryness of the air. After a successful picture, the next one would have a black line across it. A new crack had opened in some part or other of the apparatus, and had to be found out, and then stopped up with white-lead, before anything further could be attempted.

At the close of a trying day of this sort, visitors were announced; they proved to be our friends, Mr. Smith and his eldest son. Both were to start for England by the next steamer, but could not leave the island, without coming to see how we were prospering on that particular part of the mountain, which they had recommended.

What splendid carriers are the mules of Teneriffe, thought we, as Mr. Smith's animals were unloaded, and we saw how much they had brought in one day, over so many miles of lava, and up 8900 feet perpendicular. What a fuss there would have been in another country, when attempting to carry a fraction of the same weight, on the backs of Coolies or Hottentots.

With our friends contributing more than half of the feast, we made a grand dinner under the tent that day. Our table was merely a building plank, mounted on a couple of boxes, but answered perfectly well: and except on a few occasions, as after much writing, my wife never regretted that she had herself proposed decreasing the bulk of our camp equipage, by declining to adopt either actual tables or chairs.

The Peak rose grandly before our dining-room door; but Mr. Smith preferred its appearance as seen

from the " Tiro del Guanches ;" a deep defile which
from its name, was, or might have been, a place
of defence to those aborigines, if anyone else had
gone out of their way to attack, and they had chosen
to defend it.

It is situated on the great crater wall, but further
westward than Guajara.   Owing to this position, the
Peak is seen from thence with much steeper sides, than
as it now appeared to us from the south; where its true
conical form had been stretched into a ridge, lying
nearly E. and W., by the protrusion of Chajorra on
one side, and Montaña Blanco on the other.   This
last-mentioned hill, or abutment of the Peak, is of
smooth surface, and light-yellow colour; but every
here and there has exudations of red lava, which have
half-stretched, half-slobbered, down the sides like so
much treacle or hasty-pudding, and show the transverse glacier-like wrinkles in perfection.

Such common-place similes occurred to us only,
when the phenomena were viewed at a distance of
several miles. Something nobler would have been
suggested had we been closer, but might not neces-
sarily, by its mention, give the reader any better idea
of the appearance presented to our eyes.   Distance
certainly enabled us to comprehend with unexpected

clearness, the manifold history of the Peak, and its outpourings, all sub-aerial; yet how were we oppressed with immensity, on finding the stage of all this wonder-working for myriads of ages, to be itself only the floor of another crater, the great crater of "*elevation;*" something of far older date still, and on an infinitely larger scale.

Probably at work in its day under the sea, from the stratified nature of all its lavas, the general absence of pumice, and all the lighter and looser materials—this crater, with a diameter of eight miles, whose walls are like giant mountain-ranges, and whose outward slopes extend over so many thousands of square miles, that its upheaval must have sensibly altered the earth's centre of gravity,—is truly a subject to be studied by "*lunologists,*" as well as geologists.

Looking eastward beyond our ravine of the spring, an opposite headland, cut sharply through, gives an excellent section of the wall of this particular specimen of Von Buch's craters of *elevation*. Its external slope is exceedingly smooth, and moderate in angle, viz., under 12°, and it thus ascends on every side from the sea, up to its culminating line; curls over slightly there, and then at once plunges down with one or more precipices to the depth of 1800 feet, and in such a manner as to leave a vast central hollow. Sailors in their ships, might sail all round the island, and have no notion of these internal annular cliffs; nor would the natives on the coast have any idea of there being such bold features, without actually making the ascent. Yet there they are, these vertical precipices, most marked by nature of any phenomena, and contrasting so powerfully with the gentleness of all exterior slopes, that they must have some striking mechanical explanation.

Together with this difference in angle, of the two sides of the crater-wall, may be remarked the arrangement of its strata. As seen from the interior, they are accurately horizontal; but they dip down towards the sea—at angles rather less than the surface —in the side section exposed by several radial ravines, themselves mechanical witnesses of much import. The uppers tratum appears to be a bed of trachyte, some 500 feet thick, brown from oxidation, and picturesque from its weathering into angular points; underneath that is a bed of white tufa, some fifty feet thick; and this can be traced for a great distance, both in the ravine, and then round a corner, along the interior wall of the crater. Below that are other beds of trachyte and greenstone, but not so clearly and continuously seen, by reason of accumulations of *debris* lying upon them.

From details of material, rising upwards to a bird's-eye view of the whole arrangement, we find Guajara to constitute the highest portion of the huge circular rampart of this elevation volcano. It is, in fact, speaking in Vesuvian terms, the present Monte Somma of Teneriffe.

The Somma crater of Italy, was an entire ring; until broken up, on the seaward side, in the eruption of Pliny; when the cone of Vesuvius rose. Teneriffe had also its Plinian eruption, but long before the days of Rome; and raised "its Peak," or Vesuvian cone, nearly in the middle of its Sommian basin, and without destroying much of the wall. Large gaps, however, exist to the N.W. and W.; and there are such irregularities among the still remaining portions, that unless we can rise, physically or mentally, into such an elevated position, as to take a general view of the whole—we shall be but bewildered with Peaks, table-tops, and buttressing ridges of mountains. The immense scale of things before us, is exceedingly difficult to realize. The volcanic lines of Somma, must have enclosed a pretty roomy plain for Spartacus and his army of gladiators to encamp in;

but the corresponding portion of Teneriffe, which we may call the crater of Guajara, would afford space enough for the manœuvres of a whole campaign, between the largest imperial armies.

Having acquired some real idea of this immense plutonian amphitheatre, we look towards the west, and see there the steeper walls of the Caldera of Palma; eastwards, Grand Canary, an entire crater. Then remembering Von Buch's relations of terrific outbreaks in Lancerote during the last century, and flames seen rising through the sea between the islands, we cannot but look on this whole Canarian Archipelago, as constituting one enormous volcano, still to arise out of the ocean in all its majesty. While in the course of those secular changes, which Darwin has so well brought out in his researches on the Pacific,—those heavings and sinkings that the earth's crust undergoes so slowly, yet so extensively, as to have both elevated and submerged the Andes more than once,—the African continent may one day be ramparted on the west, by a greater than Andean chain of mountains; of which Madeira, the Canaries, and the Cape de Verdes, will be some of the most glorious summits.

At this period of the year, July, the sun at its set-

ting, actually went down behind the suggestive volcanic heads of Palma, and never were the forms
brought out more grandly, than on the evening of the
27th.   The air was calm, as well as transcendently
clear.   The saffron and orange of the horizon, and
rose-pink blush of the upper sky, defied artistic
imitation; while, looking down at the blue-white
rollers of the ocean of cloud below us, and the
arrowy beams of sunlight glowing along their tops,
one was reminded of a sunset amongst Polar bergs, of
nought but ice and snow.

From this scene of contemplation we were suddenly
roused by a report, that the carpenter was missing.
He had not been seen since the middle of the day,
when he had started off for a walk, no one knew
where.   He had previously tried to persuade each and
every man on the station to pedestrianize with him,
but he was the only one who cared for that active
method, of shaking off the effects of a week's hard
work; so he had started off by himself, in a new
straw hat, and a clean white jacket.

Darkness was falling so fast, that no one could with
safety go very far in any direction, and no one knew
in what quarter the wanderer was most likely to be
found; so we raised a great fire of *retama* wood on a

prominent point, and after bawling ourselves hoarse, from projecting rocks, without receiving any response, waited the approach of day, with trouble it were difficult to define.

CHAPTER V.

SOLAR RADIATION.

MORNING came, and showed us all the minutiæ of hill-tops far and near, above the sea of clouds; but no carpenter. Again we made a great fire, with green wood now, so that its white smoke rising in a high pillar, might enable the lost seaman, unskilled in wild mountain lore, to recognise the top of Guajara. Several wide circles were swept around the station, and one of the Spaniards descended into the crater, and coasted along, as it were, by the foot of Guajara, reascending on the other side; but no trace of our wanderer was found.

Had he merely miscalculated the distance, and his walking powers, the previous night, and then lain down to rest,—on finding himself benighted by the rapid closing of an almost tropical twilight,—it would have been the safest and best course to pursue; for, in darkness, a precipice might easily be stumbled over; while on the other hand, in this warm summer

M

season, and in a land without large wild animals of any description, a man could take no harm during a night's repose under a sheltering rock. But then thought we, in such case he would surely have started again soon after day-break, and would have returned to the tents at an early hour. Here however, eleven o'clock in the forenoon had arrived, and there was no appearance of him yet.

Then a fearful idea occurred, of the man having fallen over a precipice; and either being killed outright, or lying amongst the stones wounded, maimed, and unable to give, or return, any signal. While we were almost wishing for African vultures, to point out by their flight the quarter of the catastrophe; behold a crow, cawing sinisterly, flew above our heads, and went down over the cliff some little distance to the East; then a few minutes after, went another;—so in the absence of any more positive indication, I immediately followed their lead on foot.

The descent was not very difficult, for at one point the wall of rock had been broken up, and was lying in a long slope of stones, really awful to contemplate. It must have required some extraordinary exertions, even of nature, to . have tossed about such immense masses, in the reckless way in which she

had apparently done here; and, as it seemed, only
yesterday, so fresh and clean were all the blocks.
Availing myself of many such breaks, which appeared here and there in the ledges of precipice,
I descended several hundred feet, and in one part
found a man's footsteps; they led near, but not
exactly to, a very treacherous slope of fine *débris*,
which seemed at the commencement, to invite a
confiding descent; but presently broke off in a vertical cliff; and the next objects one could discern
beyond its upper edge, were the tops of some remarkable pinnacles of rock near the base of the mountain.
At this place I shouted long and loud; there was
no human answer; and no crows flew up, startled
from an unhallowed feast.

Pushing along by every possible path, I came at
length vertically under our station, but at a great
depth below. Here was the finest plunge of rock
in the whole mountain, in parts rather overhanging,
and not very encouraging to one in the position I
occupied; for a large slice of the front of the precipice seemed to have separated from the mass, and to
have broken up; but—the portions having caught for
the moment and held each other entangled by friction,—did not fall. There however they were, hang-

M 2

ing together only by their points; and waiting for
the smallest additional force to bring them thundering down. The carpenter was surely too clever a
fellow to go and tumble over a precipice, but how
easily might a stone have fallen from that impending
avalanche, and have smote to sudden death, either
him, or the most experienced mountaineer. So at least
I thought, on gazing up at the promised ruin, when
suddenly there was a step close behind me. Starting
round, there was a black goat. After looking at each
other for a little, the animal gave a pulling sort
of bite at a branch of *retama*, and then hopped and
jumped about amongst the blocks of stone, until it
came to another bush. By and by a second goat
appeared; a white one this, and all that could be
made out from the movements of the pair, was, that
they were by no means afraid of mankind.

Coming again to the station, and finding that with
the exception of three more crows, which had been
seen flying in an opposite direction to the two preceding ones, no sign of any sort had appeared; and
being convinced from the little specimen I had been
having of the vast extent, complications, and difficulties of our Mount Guajara,—that a dead or badly
wounded man might lie for weeks unfound by a

few searchers,—I sent off a deputation to ask assistance from the alcalde of the village of Chasna. The
loss of a man on the mountain, our interpreter assured
us, was by no means an uncommon event, and the
country people were quite accustomed to being called
out, to assist in a finding.

Our deputation had been gone some hours, when a
sort of spectre appeared; it was the carpenter himself,
looking strangely wild and haggard, not altogether
fully sensible, and with the soles nearly worn off his
boots. He was conducted by an old goat-herd, and his
story was this. On setting off for his solitary walk the
previous day, he had gone down to the spring, indulged in a good drink, and a sleep; after which he
looked about, and saw the Peak so wonderfully close,
that he thought he would just step down the side of the
hill, and then cross over the crater floor, to where a
very black stream of lava, obsidian he hoped, descended
from the cone. The mere stepping down, occupied
much longer than he had expected; and he had hardly
begun to cross the plain, ere he was bewildered amongst
heaps and heaps of lava stones, rising like hills above
his head, and shutting out all the distant view. He
attempted when too late to retrace his steps, but
taking the wrong direction, did not extricate himself

from the stony wilderness, before darkness compelled him to stop. Even next day, though he did regain the Canadas, he had so lost all idea of the lay of the country, and all his strength from hunger and thirst, that if the old goat-herd had not met him, and given him milk and rest for a while in his hut, and then brought him up to us, he would certainly have been a lost carpenter.

The countryman was a fine hale fellow, and a splendid walker, though his hair was grey. But his *morale*, was finer even than his *physique;* for he exclaimed violently against the idea of taking any money for what he had done; and though he had no objection to breaking his fast with us, and even carrying home a few biscuits; still he insisted on giving us, for them, some milk out of a goatskin which he carried over his shoulder.

On subsequent occasions we frequently visited with photographic camera, the scenes of cleft and shattered cliffs with which our day of search had made us acquainted. The dangerous front of precipice below the station, was a very old one, as evidenced by the discoloration from weathering and lichens, on its face; and as it might apparently fall at any moment, there came the interesting inquiry, whether, when that event did occur, the thickness of slice taken off, would not reach backward as far, or farther, than our tents on the summit. Some actually fallen portions in the neighbourhood, must have been ruins of very recent date, the angles were so perfectly sharp, and the surfaces looked so new; and were as brilliant with their embedded crystals of felspar, as any fresh fractures we could make with the hammer. In some places where large parts of the trachyte cliff had been thus destroyed, the dykes having been broken up, together with the matrix, caused curiously stray blocks of greenstone, and of a black lava to be found. (*See Photo-stereograph, No.* 8.) This dark stone attracted most attention, being so utterly dissimilar to the lighter and gayer coloured trachytes of Guajara; but we were afterwards to become acquainted with it, as forming the whole material of the more recent portion of the Peak, or cone of eruption.

The first of August was again a term-day of hourly meteorological observations, and was to be kept by Captain L. Corke in the yacht Titania, as well as by ourselves; so that there might be exact comparison between the climate below and aloft. The arrangement had been communicated by our friend Mr.

Hamilton in Santa Cruz, who writing to us soon after, said that he had no doubt Captain Corke had observed throughout the desired twenty-four hours, for they saw nothing of him on shore during the whole of the first and second of the month. So it turned out too, for the excellent seaman, both upon this and subsequent occasions, himself made every observation that was required, during both day and night.

On Guajara, there was no fear of oversleeping ourselves on the morning of the first; for from an early hour, the wind rained alarums upon us, in the shape of gusts and eddies that made our tent reel again, and caused its pole to swing about like the mast of some little boat in a cross sea. The character of the wind, which was from the N.E., was remarkable; there was never less, than a velocity of some fifteen miles an hour, in the aerial current; while every now and then came blasts of far greater force. We first heard them in the distance, as they struck against the face of the precipice below, and seemed to roar with rage; and then immediately came whistling over the cliffs, and caught our poor tent on its only undefended side; viz., the northern one, left purposely open that we might enjoy that splendid feature of the landscape— the mighty Peak.

What with the strength and chilliness of the wind, though its temperature was 57°, the poor Spaniards looked like drowned rats, and about as miserable, cowering in their long cloaks, under rocks and stones for shelter. From some experiments which I made with the sailors' assistance, the velocity of wind at 5 A.M. was about 30 miles an hour, but fortunately decreased to about a third of that before the middle of the day. During all the period of greatest violence, the usual sea of N.E. cloud, was at its ordinary elevation of only from 3000 to 5000 feet, and there were no upper clouds; yet here was the N.E. current of wind felt, and in such intensity, at an elevation of 8900 feet; and how much higher, there was no saying. Above all this Polar influence, there was no doubt prevailing a legitimate return current from the equator; but still the affair of this morning most eminently confirmed the general result of previous days, viz., that the N.E. cloud is found low down in its N.E. stratum of wind; and not, according to the general belief, midway between the N.E. and S.W. currents.

That midway position appeared, from a summing of all our experiences, to be, in the summer time at least of these latitudes, a region of calm as to wind, and of freedom from any sort of cloud. As the depth of lower N.E. current rose or fell, so did this zone of calm, which was only arrived at gradually, by the air in motion, above and below. The line of separation therefore, must have fallen greatly, perhaps three thousand feet, on the morning when we had the S.W. storm; and now, with a similar visitation from the N.E., there must have been an equivalent rise. Hence we may at once see the futility of certain proposals, to determine by one or two ascents, the exact height of separation between the Polar and Equatorial currents, in a given latitude. If a few days, had produced variations of several thousand feet; a few months, with their effects of season, would bring about far more change; and so indeed we found before we had finished with our expedition, when the S.W. wind descended to the very surface of the sea.

Amongst other instruments which we had prepared for observation on this day, were two large black bulb thermometers, kindly lent by Mr. Airy, and a smaller one by Dr. Lee, for measuring solar radiation, one of the most characteristic features of a mountain climate. Readings were taken every five minutes during the greater part of the day; and on subse-

quent occasions, even at one minute intervals, the
changes were so rapid.   At this work I should soon
have been exhausted, had not the second mate of the
yacht shown so much taste and talent for observation,
as to be able, after a very little instruction, to note
accurately the indications of many instruments, which
he had never seen or heard of before.

With this assistance, a larger series of radiation
measures, and with higher results, was procured, than
had perhaps ever been taken before.   On the first
day of trial, the sun thermometer rose so rapidly, that
before we knew what we were about, Dr. Lee's instrument, a " patent " one for this sort of observation, by
a London maker, and with its tube extending to
140°, was broken to pieces by the mercury passing
that temperature.   Mr. Airy's instruments however,
not only were graduated to 178° and 180°, but had a
bulb at the top of the tube, into which the mercury
could pass with safety at higher temperatures; and
with these we had the satisfaction of seeing the
quicksilver stand at 168° at noon, even on our windy
term-day; when the temperature was only 67°.

High though this result might be, it was far excelled on subsequent days, when the calm atmosphere
being more favourable to obtaining a true result, our

exposed thermometer rose to 180° by half-past nine
o'clock in the morning; and at 12 o'clock, had half
filled its spare bulb. What then, the reader may
ask, was the maximum heat of radiation on a fair
day? Why, although the insufficiency of the instruments prevented our ever actually seeing it there, yet
from a comparison of curves, whole or partial, on
many different days, it results that on August 4th,
the black-bulb temperature in the sun must have
been 212°. 4, the temperature in the shade being only
60°; thus leaving the enormous quantity of 152° for
the effect of sunshine at a height of 8900 feet.

To chronicle the exact circumstances under which
these high results appeared, the photographic camera
was employed with effect. Accordingly in *Photo-stereograph*, No. 7, may be seen, towards one corner of the
telescope enclosure, our stout seaman-observer, note-
book in one hand, and chronometer in the other,
counting seconds up to the moment that he is to take
the reading of the exposed thermometer, sharp: after
that, he will remove part of the tin-foil covered lid from
the sheltered instrument, in order to get the temperature of shade. Both thermometers have their
bulbs encased in glass bells, from which the air has
been extracted by syringes, that project through the

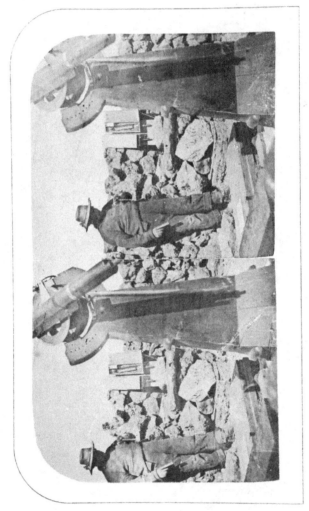

SECOND MATE OF YACHT OBSERVING RADIATION THERMOMETERS ON MOUNT GUAJARA.

...

Printed by A. J. Millhush, under the superintendence of James Glaisher, Esq. F. R. S.
and published by Lovell Reeve.

boxes below; and show their turned rings neatly
under a magnifying glass.   Our honest second mate
wants no such refinement of method to make him
visible; and though he had requested that his portrait
might be taken,—in the act of holding up a large sextant, which he was ambitioning to learn the use of; and
with a smart cap on his head, and in his best jacket, as
if he were already a merchant skipper of some degree,
—I preferred catching him at an instant when he was
thinking of nothing but his duty; with his oldest
Guernsey on his broad, manly breast; and his trousers
turned up and dusty, from his recent labours at the
wall.

Amongst those who have not thought much on radiation, there has arisen an idea, that the closer we go
to the sun, the less heat do we receive from that great
luminary. Consequently, they allege, a comet—which
like that of 1843 passed at its perihelion within 60,000
miles of the surface of the sun,—must have suffered
rather from cold, the cold of space, than from the excessive heat which astronomers have described; and
the circumstance of perpetual snow on the top of our
loftiest mountains, is pointed to as an illustration of
the case.   Such snow however, and its perpetuity, are
rather to be taken as proofs of temperature, and of the

non-conductive powers of frozen water, than of radiation; and this latter quality we invariably found on Guajara to be the opposite of temperature; *i.e.* the days in which the temperature in the shade was lowest, and we may add, when the air was most clear and transparent—were precisely the days of highest radiation. With the increase, on the other hand, of our too frequent visitors, the banks of dust-haze, the height of a thermometer in the shade increased; while another exposed to the sun's rays, invariably decreased.

Were it possible to ascend fifty miles above the earth's surface, and so escape altogether from its atmosphere, the temperature of shade might fall to —50°, but that of solar exposure, or radiation, would be increased probably some hundreds of degrees, though we had not sensibly altered our distance from the sun. Were we then, from that point, enabled to accompany a comet on its perihelion approach to the mighty orb of light and heat, the immediate radiation effect would be continually increasing, according to the greater angular area under which the sun was seen; while the resultant effect made good on any material body, would be greater and greater according to the increased proportion that the area of sun surface bore to the extent of visible sky. At our distance from the sun, this body is at a disadvantage compared to the sky and earth, of about $\frac{1}{185,000}$th; but at the distance at which some comets have been, when the sun's diameter must have appeared to them under very nearly an angle of 180°, the hot and cold surfaces would be reduced to equality; and the rarity therefore of the ethereal medium pervading space, and its low temperature, would by no means suffice to cool down one side of the comet, as fast as the magnified sun roasted up the other.

Amongst those who *have* thought on, and observed the radiation of the heavenly bodies, whether accurately as Saussure in his ascent of Mont Blanc, Sabine, Mason, Herschel, and Daniell; or, with reference only to general impressions, as much earlier travellers—there have been truer ideas of the increase of solar radiation, corresponding to every ascent in the atmosphere. Something of the sort must even have dictated that eastern tale, still in vogue amongst Egyptians, touching King Solomon and the vultures. That incomparable king, wishing to visit some distant part of the earth, summoned his obedient Jinn to convey him through the air to the place in question. Whereupon those spiritual giants, seizing the four corners of the carpet on which their monarch was seated,

raised it high in air, above all the clouds, and then
flew along with it towards the spot appointed. They
flew swiftly, as only Jinn can fly, and everything went
on as right as could be, until the king felt the radiation of the sun so strong, that he asked a flock of
vultures, whom he passed in upper air, to come and
fly between him and that blazing orb, whose rays
were beating down so unmercifully on his kingly
head and shoulders. What the vultures answered the
king, and what the king said to them again, and what
he did after that, being more interesting to naturalists
than to astronomers, as offering an explanation of
vulturian neck-nakedness, I shall not stay to describe ;
suffice it that the increased radiation above the clouds,
was thought powerful enough to disturb the equa-
nimity of a sovereign of Palestine, a country where,
from the clearness of its atmosphere, the natives must
be accustomed to a very considerable intensity of
solar rays, even in their lowest valleys.

The suddenness with which day changes into night
on high mountains, has often been remarked ; and as
the sun went down on Guajara, we found radiation
decrease, with marked rapidity ; the amount fallen
through in the last hour, being no less than 82°. At
sunrise during the same period the rise had been 69° ;

with the remarkable feature, that at sunrise itself, the
radiation was negative; the exposed thermometer
reading less than the shaded one, by 4°. This circumstance was but in confirmation of that weakness of the
sun at rising, which we had generally remarked. He
was indeed very unlike the South African globe, of
which Cape boers relate, " So soon, Mynheer, as the
sun comes up, that instant he sticks you through."
We ought however to bear in mind, that on a very
high mountain, the *rising* sun is seen through a
greater thickness of the atmosphere, than he can ever
be on a level plain; and as, moreover, he rose on
Guajara at a zenith distance of 91° 10', all his shining
from that instant, until he had reached 90°, was so
much given in, over and above, to mountain climate.
Besides this, when he had attained the altitude at
which he is usually first seen by dwellers below, he
had quite got over his early indecision; and in an
hour had heated up an exposed thermometer to 112°.

At night, the radiation was constantly negative,
and on one occasion, a thermometer exposed to the
sky, was 17° below a shaded bulb; and this having
itself fallen 16° beneath its day maximum, we have a
difference in temperature for an exposed body, of 185°
between noon and night; while if to this we add the

N

effect of evaporation at the coldest period, the difference of wet and dry bulb thermometers often reaching
to more than 20°, there will be a fearful amount of
change for the human constitution to withstand. We
all had nevertheless most excellent health; and towards the small hours of morning on the term-day,
when I might have retired to the tent, I preferred
remaining out in the calm, open air, gazing at an
extraordinary cloud suspended immediately over our
station.

First attracting attention from a slight-luminosity
at 3h. 43m. A.M.,—when it could hardly have been
illumined by the sun, unless we suppose a height incompatible with subsequent appearances,—this cloud
became presently more remarkable for the changes of
form, and we may say of nature, which it underwent;
and without any alteration of place. Hour after hour,
and far into the next day, did this cloud remain
vertically over our heads; yet it was unconnected
with our mountain-top, for there was nothing similar
over any of the other heads of the crater wall, nor
over the loftier central Peak; but there were two over
the sea, or rather the cloud-sea; one to the south,
and another to the east; both of them, as it appeared
to us, permanently located there. Yet with all this

constancy in general position, the clouds were alter-
ing their figures from minute to minute, and even at
every second.

In the dark, we could see the zenith mass by its
luminous fringe, changing form fitfully; now resolving
itself into a globe, then into two, and again into one;
and when twilight came, and when other clouds were
of brown and blue greys, our particular one was
intensely white, and seemed to be bent on rivalling
flashes of aurora, or the changes of artificial fireworks. There was a nimbus portion, and this threw
out curving bands of cirri around it, then chased
them round and round, as a kitten does its tail, and
proceeded still further, to swallow them; growing as
it did so, and becoming a most portentous-looking
cloud. Presently again the nimbus arranged itself in
a series of many lenticular discs, one above the other,
as in Rhumkorff's electric discharge; then throwing
out auroral striæ, it collapsed once more into a single
mass; and all this in the course of a few seconds.

CHAPTER VI.

WHIRLWINDS, AND VISITORS.

THE N.E. storm of the 1st of August, though exhausted of its greatest force in a few hours, was still dragging its slow length along through the following day. Every now and then a stray puff of wind came hurrying up, like a conspirator too late for the concerted rising, and throwing itself wildly against the northern cliffs of Guajara, came spinning over their summits in the form of a little whirlwind. Sometimes, even our stone walls were enough to produce the effect; and then we saw an ordinary gust of wind blowing over the dry ground outside, raising a cloud of sand before it, and at the instant that it struck our wall, it eddied round the corner in a fine little revolving pillar of dust, occasionally entering our tent door in a most inconvenient manner. Amongst numerous young hurricanes which were thus generated before our eyes, one of them was remarkable for a regular increase in its diameter, as it travelled on, in a curving line; and for its diminishing velocity at the same time in the whirl. Mr. Redfield's theory, and Col. Reid's plates, of the West Indian storms, could not have been better illustrated.

Occasionally something more powerful came by; and one day, a heavy piece of canvas, ten feet square, spread out on the rock, was suddenly lifted up, whirled round and round in a horizontal plane, and then deposited again, as flat as before, almost in its former position.

On another occasion we had a more curious display still, of what the wind could do in goodnatured sport. Being at work inside the telescope enclosure, I felt myself suddenly peppered with little pebbles, coming from over the wall; and jumping up to see what this meant, I was almost blinded by a cloud of dust, and a confused blast of wind; in the midst of which there was an immense rustling noise. This had been caused by one of the last things which we had purchased before leaving England, no less than a draper's whole roll of cheap blue cotton cloth. On this very morning we had opened the box in which the material was packed, to take a portion for some purpose; and the chest was still open, with a corner of the calico projecting. What then did the

wind do, but seize hold of an end of the cloth, unroll
the whole of it, and carry it clean out of the enclosure. As soon as I could open my eyes, lo! there
was our piece of blue cloth, forty yards long, right
up in the sky, and projected most intensely against
some white cirrous clouds. So high was it, that it
looked like a mere piece of ribbon. Three times completely did it sail slowly round in a circle—accompanied by some hats, caps, and other smaller matters,
that looked like swallows beside it—and then descending leisurely, it fell about four hundred yards
to the S.W. of our position.

There was no doubt about the direction of the
path of this whirlwind, being at first, nearly identical
with that of its parent wind; and I never had any
doubt when looking at it, that the direction of whirl
was similar to that of the hands of a watch: but such
an observation requires perhaps more experience to be
unexceptionable.

The direction of whirl is of extreme importance to
the true explanation of hurricanes; concerning which
some most instructive hints may be derived from
these Guajara experiences. For on considering the
mighty amount of mechanical power set in action by
the Trade-winds, during the long course of their sway

over the ocean; and then the sudden and abnormal manner in which they are interfered with, by portions of land stretching across the tropical seas,—there is every reason to expect that some disturbances in the nature of eddies must be produced, similar in principle to those about our station. We ought also on mechanical grounds, to find them in the North and South Atlantic, not towards their Eastern, but rather their Western sides; or exactly where the cyclones of the West Indies, and of Rio Janeiro are known to prevail. Of these storms the dynamic force is such, that we can only look to the Trade-wind current, which we know has been interrupted, for a sufficient producing power; and the interference becomes more complicated, from the now apparently proved idea, that those Polar currents in either hemisphere, cross over each other at the Equator; so that West Indian hurricanes, may be due in some measure, to descending eddies of a S.E. Trade-wind.

Electrical hypotheses have been started by some persons to explain revolving storms; but while it would be a difficult matter for any one to produce enough electricity, to supply the force exerted, even in the little affair of carrying off our piece of calico—if they did at last succeed therein, the mechanical means taken up for the purpose, and which must be accounted for in nature, could certainly be employed more economically in bringing about the observed result, directly. The whirlwinds on Guajara, small as they were, required an exertion of a considerable amount of mechanical energy; and the question to be settled, is, which of the two elements present on the occasion, was most capable of producing the effect observed, wind or electricity? Of the vigour of the wind there was no doubt to any one who stood out in it, or even to one in the tent that was being shaken so rudely; but who will vouch for the amount of electricity in the air being sufficient for the purpose? This amount must have been exceedingly small; as, notwithstanding the dryness of the atmosphere, there was no crackling on rubbing silk, and there were no sparks on stroking fur; and even with the delicate test of the gold-leaf strips of our electrometer, there was barely power to separate them; this too, nearly equally, whether plain gusts or whirls were passing by.

In short, whatever the N.E. wind did, its electricity was always moderate. Sometimes we saw its cloud-sea, especially in the neighbourhood of Gomera, gradually changing from cumuloni to cumuli, and almost to lightning cumuli. But there was still no electric demonstration, and a fine specimen of the "Table Cloth," or "*Perruque*" cloud of Table Mountain, combed as it were smoothly down over the highest part of the island, was all that rewarded our watching. Sometimes again, the S.W. clouds floating high over our heads, looked very like cirro-cumulo-strati; and I anxiously cast about for some method of lightning protection, and looked with awe at the number of round holes formed in, or as some would at once have said, struck into, and through, the rocks around. The influences of such clouds, however, did not penetrate down to us, or our electrometer; and when they had sailed away to spend their energies in another land, we ascertained to our satisfaction, that the round holes were due to a special manner of formation and decomposition of the trachytic lava.

The tract of sea between Gomera and Palma, next to that on the N.E. of Teneriffe, seemed to be a special arena for cloud manifestations. Generally there was a tumultuous assemblage of rollers at the height of about 4000 feet. But on the evening of the 3rd of August, these were all cleared away, and presently, invisible hands seemed to begin the laying down of a new stratum, which came out dense and strong from behind Chajorra, and stretched away into distance, preserving a level of 5000 feet; a strange stage, suspended in mid-air. Then appeared a number of fleecy little wisps of white cloud, that danced before us, like aerial sprites, on the dark grey platform of cumuloni, until long after the sun went down.

Next day, though one could not pretend that the air had become damp, its dryness had decreased to a dew-point depression of 18°, and some of those aerial sprites of the previous evening, paid us a visit at noon-day. Without the level of the Trade-wind cloud itself being altered, little puffs of cirrous mist on the top of it, rose up and passed over our heads from the S.E.; then others came, and for a few minutes enveloped us with fog; but they disappeared again immediately, and the sky became bluer, and the sun brighter, than ever.

Trying the capacity of our telescope to pick up stars by daylight, I found it able to do far more than in Edinburgh, though its superiority was not so decidedly predominant as at night, owing apparently to the increased power of the sun. The atmosphere was far clearer and purer than down below; but then the sun was to such an extent brighter, as to illumine the atmospheric motes to a degree, that quite precluded a chance of seeing anything in his immediate neighbourhood. Hence perhaps it came about, that the chief fact impressed on us by day observations, was, the very great preponderance in brightness of large stars over small ones. Sirius for instance blazed in the telescope at mid-day with a lustre almost painful, while smaller stars of the first magnitude were not very notable, and total invisibility arrived with those of the third magnitude.

The sun himself was of course frequently looked at; but was never so well defined as the stars were; his tremendous radiation doubtless perverting the telescope, as well as disturbing the air. Added to this, the year being nearly at the minimum of Solar spots, there was seldom anything very notable in progress, on that wondrous disc of light. Various devices were employed from time to time, to make the "eclipse red prominences" visible, if in existence; but invariably without success.

In the afternoon came a number of peasants, fine specimens of the country people, clean, obliging, and cheerful. There were several men, one woman, and a

number of boys, who—in their simple dress of white
shirt and drawers, tied with a scarlet band about their
waists, a broad hat, and nothing besides save their
fine coffee complexion, black beaded eyes, and teeth
whiter than ivory,—formed most Murillo-like groups,
as they played about amongst the rocks, with diminutive iron-shod poles; and then lay panting and
smiling in the shade.

The elders of this party had come for something
more important; for they had heard rumours in their
goat-tending country, that an English Astronomer
had arrived with so large a telescope, that he had
actually been able to see goats in the moon. No
civilization is possible, or is attempted in Teneriffe
without a certain understratum of these animals,
which furnish nearly all that the simple natives require for food and drink, shelter and clothing. From
thence arose a notion amongst them, that Lunarians
cannot carry on the business of life, in any other
way or by any other means. In their inability to
separate the idea of a rational existence from human
beings in general, and themselves in particular, these
honest goat-herds have but too many to keep them in
countenance in our own country.

The visitors were however shown other things that

astonished them greatly, and perhaps they admired nothing more, than the stupendous size and fresh-coloured complexion of one of our sailors. While the woman of their party, being taken by my wife into her tent, touched everything respectfully with her fingers as she looked at it; and con-cluded with a congratulation to herself, that she had been so courageous as to leave her goat-fold, and ascend the mountain-top, to see such wonders as these.

Though the "Islenos" pass as Spaniards, one cannot but confess, when wandering in Teneriffe with perfect safety, through most guerilla-looking passes—that there is on the whole a certain difference. When for instance, a murder was committed in the island some years since, everyone immediately de-clared that it could not have been the work of a native; and sure enough the perpetrator was soon apprehended, and found to be a disbanded soldier from Spain. While too, much might be concluded from the prevalence still of certain Guanche customs, and the use of many of their words,—more impor-tance may be attached, it is to be hoped for hu-manity's sake, to the mild modification that Spanish character has evidently undergone here; seeing that the effect superinduced, is precisely in accordance with what history has handed down to us, respecting the aborigines of these islands.

First discovered to Europe in 1330, and then assailed, invaded, and re-invaded by Normans, Portuguese and Spaniards; bravely defending their homes, defeating over and over again their outnumbering enemies, clad in armour and furnished with weapons of steel, but treating their captives always with clemency and more mercy than they deserved,—yet tired out at last, and overwhelmed by the ceaseless shoals of aggressors poured upon them from the millions of Spain—the brave Guanches of Grand Canary and Teneriffe at length gave in after a century of fighting; the smaller islands had been subdued before. Had they possessed a large country, these men might still have preserved their independence; but as it was, a longer persistence in the conflict could but have ended, in their total extermination. Having shown what they could do as men, in bravery and mercy, they yielded to natural laws which had suddenly brought a great, a civilized, and a warlike as well as aggressive nation, into contact with them, a tribe of primitive goatherds, on a finite tract of some few dozen miles.

The best English account of these Guanches is in Captain Glas's early, but excellent work. He dissents from the common idea of extermination, and considers that this interesting race, has rather passed into the bulk of the Spanish people, adopting their name and language. The present Canarian goats are allowed on all hands to be descended from aboriginal flocks. Real Guanche too, both in name and manner of composition, is the " gofio ;" a staple food with all Teneriffian peasants; and which our friendly visitors began shortly to prepare for themselves.

Some toasted meal, now of Indian corn, is put into a kid-skin prepared with all the legs dangling about it, in true southern style. Some water being added, and the bag's mouth, the quondam neck of the little kid, being tied up fast, a man sets to, energetically kneading up the skin upon a flat stone; whereupon, with the meal and water inside, the skin-bag tosses out its four legs violently, and appears to be taken with terrible convulsions. After a while the bag is opened, the paste that has been formed is taken out as good " gofio," and is eaten without further ceremony. During winter a little milk, and a flavour of cheese are added; and so good is the dish thought to be, in spite of its simplicity, and want of cooking, that it forms the children's breakfast in many of the best families of Canaria.

The goat-herd party having hastened off, when the sun became low; and the air being still transcendently clear, I prepared for a hard telescopic night; and remained at it from dark to half-past two o'clock in the morning. The negative radiation had amounted to 18°, and the dryness had increased again to a dew-point depression of 30°; but the air was calm, and the stars were so vividly bright, as well as admirably defined, that I had no sense either of fatigue, or of cold. In Edinburgh, the telescope which I was using, had never exhibited satisfactorily the companion of α Lyræ, a star of the eleventh magnitude; but now this little speck of light was shown so brilliantly, that I could hardly but suspect some alteration in the star, until I had looked at the companion of δ Equulei, also of the eleventh magnitude, and found it as bright and distinct. Then trying further; B of δ Aquilæ, C of 185 Antinoi, and D of 13 Lyræ, stars of the twelfth magnitude, were seen: also B of 128 Anseris, B of 307 Aquilæ, both of the thirteenth magnitude; lastly even C of 5 Aquilæ, and C of β Equulei which are of no more than the fourteenth magnitude.

TRACHYTE BLOCKS ON GUAJARA.

p. 167.

Printed by A. I. Melhuish under the superintendence of James Glaisher Esq.ʳ F.R.S
and published by Lovell Reeve.

Here was indeed an extraordinary increase of the space penetrating power of this telescope, viz. from the tenth to the fourteenth magnitude. The stars were not only so much brighter, but also so much better defined than I had been accustomed to at the sea level, that they appeared to provoke one to apply micrometer measurement to them. Had this been the highest point of Teneriffe, then here would I immediately have tried to bring up the Pattinson telescope. But as long as there were more elevated points still, every success on Guajara, was only an additional reason for trying such higher levels.

In the numerous consultations which I had held on the subject of those more elevated regions of craters and lava streams, there was so much more of hearsay than actual knowledge,—that not being able, myself, to leave Guajara, I at last sent off the interpreter and a guide, to make a systematic exploration of that upper world.

On the very night after they had left, a fire was seen on the Peak. It was too soon to be our party; who could it be? Our native assistants were in extraordinary ferment at this sign of other life and activity on the heights, than ourselves. The fire was

o

about a thousand feet above the level of Guajara, and on looking at it through the telescope—we saw tongues of flame leaping fitfully amongst black rocks, while at intervals the forms of men were revealed for a moment by lurid flashes, and then all disappeared in gloom.

CHAPTER VII.

DROUGHT AND LIGHT.

ABOUT a mile to the south of our station, along the descending slope, was a curving ridge of rocks, the mouldering rim of a small crater; and interesting, from the rare feature of verticality in some parts of its exterior. Just in front of it, was apparently, a road across the mountain; the "road of the Guanche kings," we felt inclined to name it. Not a bush or rock was there visible, and the sides were bordered or marked out, by large blocks at short intervals. But then who would make so grand a line for traffic through desolate parts of the mountain, when there were only mule paths in the cultivated regions; a line too, from 100 to 150 feet broad, and leading only from a precipice on the west, over a barren and stony ridge, to another precipice on the east?

We went down to examine the place; and as soon as we arrived on the seeming road—a tract sure enough cleared of all large obstacles—we at once

began to sink in over our ankles, in a gravel of yellow
and brown stones, of remarkably small specific gravity,
and acting almost like a quicksand. The change was
startling, after the rock-firm surface of Guajara; and
tracing the line both ways, we found neither more
nor less, than these symptoms of a danger of sinking in.

According to Humboldt, there are well-attested
instances of trachyte mountains having suddenly
opened, vomiting forth volcanic fire for a time, and
then closing up for ever,—surely then, this rent across
Guajara, imperfectly filled with ashes and pumice,
might be the *locale* of such an event.

Other explanations also occurred, as worthy of con-
sideration; but with the small crater invitingly close,
we entered by a convenient gap, and found the interior almost silted up from the seaward side; where
its walls had been washed to such an extent, as to
have nearly disappeared. The finest remains of this
piece of antiquity, for it must certainly date from the
submarine period, are towards the west. Following
this portion of the arc, we came to an aperture
from which there had been a distinct eruption, and
an outflow of a stream of dense obsidian, alternately
laminated with green felspar, coated reddish brown

externally. Round about were lumps of the purest and densest obsidian, that are to be met with in the whole island. Easily breaking into thin edges, with which deal could be quickly sliced away, we were quite able to understand the importance of this volcanic glass amongst the poor Guanches in former days, ignorant of any species of metal.

Returning up to our station in a zigzag direction, we came on many small lines of rocky wall, of which there are better instances still, on the westward heights. They seem to indicate, that in its uprise from the ocean, the great crater for a long while had little more than its lip above the waves; which then cut back into the sloping sides, and formed these quondam sea-cliffs. The smoothness of the remaining portion of descent, covered as it is with soil, indicates for that part a quicker rise, or a more quiet sea. While the high cliffs, now formed in a great measure along the coasts,—and evidently broken and washed out by waves, on the same principle as that recent scene of destruction, west of Orotava, described in Chapter IV. Part 1; but requiring myriads of similar storms to produce the whole effect,—may be held as proving that the present rate of ascent is extremely slow.

Specimens of Guajara trachyte, we found on examination to be not only magnetic, but to have polarity;
while the laminated mixture from the small crater,
was more eminent still in its action.

On taking out the apparatus for these experiments,
it was found to be suffering from the same intense
drought, under which most of our instruments, or their
boxes, were giving way, and which seemed to arrive at
its maximum in the beginning of August. A wooden-
scaled thermometer had then twisted or started into
such a curve, as not only to break the glass stem, but
to eject the central portion to a considerable distance.
Smart mahogany cases had cracks in their lids, into
which you might put your finger; and lifting up a box
carefully by both handles, I raised only its lid and sides.
The glue, fastening the bottom and all the partitions
and lockers, had given way completely; over aridity,
acting like damp, in destroying the cohesive power of
the animal matter.

Again the cork of a bottle of Canada balsam had
shrunk to such an extent, as to let out the sticky
fluid amongst microscopical glasses, and glue them
all into one odious lump. Worse still, the wooden
base of our electrometer, in shrinking on the glass
bell, had broken it; and finally the well-seasoned

mahogany box of the magnetometer had contracted on its plate-glass cover, and forcibly held it in. Fortunately the glass, being half an inch thick, resisted the pressure, until we could come to its assistance with chisels and pen-knives, and cut it out safely from so murderous a grip.

These little accidents had their scientific uses; for our wet bulb thermometer only gave the measure of dryness at the instant of observation; while the effects on our baggage, acted, though vulgarly, the part of a cumulative instrument, and indicated the sum total of drought for the entire period of our mountain residence.

The observations for intensity of terrestrial magnetism were not very satisfactory, owing to exceptional characters in the soil. We turned therefore with more hope to optical questions, wherein the height on which we were placed, was everything, and the nature of the ground nothing. Foremost amongst these, came the subject of black lines in the spectrum; primarily discovered by Wollaston, and secondly, but quite independently by Fraunhofer, and so much better taken account of practically by him, as to be now generally known by his name. First viewed as so many defects or originally

missing portions of the sun's light, suspicions were
afterwards raised as to their being caused by the
absorption of our own atmosphere.   But in that case,
even allowing for a moment such cosmical egotism, as
that the light of our own luminary is perfect in itself,
and the earth's atmosphere the sole cause of evil,—
how comes it that the light of stars, some of them
many times larger than our own sun, on passing
through the same atmosphere, presents in each of
their spectra a different series of black lines ?   The
case being altogether one of practice and experiment,
the elimination of atmospheric effects, in part at least,
by comparing observations on a high mountain with
those taken at a lower level, becomes a matter of the
first importance.

In such experiments, a dark room is generally a
standard requisite.   This was however more easily
talked about, than procured, on the mountain.   We
built the walls of an " optical" room certainly, and
roofed it in with deal planks and canvas; but the wood
was quite translucent in a vertical sun, and the canvas
appeared a mere netting; while our thick stone
walls showed an infinity of holes, through which the
light came streaming in, either from bright sky or
still brighter ground.   At last, however, what with

heaping on half a stack of *retama* branches above, and lining the inside of the room with that blue cloth, which the whirlwind had so obligingly returned to us,—a sufficient degree of darkness was procured, and the observations were begun; the solar light being reflected into the room through a narrow slit, from a mirror worked by a man outside.

At mid-day, there was not any very great difference between the system of lines which we saw, and the received estimation of them; but as the sun approached the horizon, they grew in numbers, thickness, and definition, in the most extraordinary manner. Careful drawings made both morning and evening, at length satisfactorily demonstrated, from the variations of some lines, and the constancy of others,—that there are certain of them that are produced by our own atmosphere, and others by some medium very much more distant, and probably depending on the nature of solar light.

The noting of these lines was a very long and tedious process, and I sought the assistance of photography in vain; for no exposure of the collodion plate seemed to bring on anything but a faint coloration in the region even of violet rays. But towards the end of the time, trying the same experiment

with one of his quartz trains, obligingly lent by Professor Stokes himself, where the rays of light passed
through no glass medium,—instantly the photographic action was most intense.   One minute, ten
seconds, five seconds, one second, were tried as intervals, and still the plates were solarized.   At length
with a contrivance to reduce the exposure to a tenth
of a second,—a fine picture was produced; and lines
were mapped down with more certainty in that short
time, than in days by eye and pencil.

With great ideas of what might be done by
means of a quartz objective in a camera, and hopes
of better opportunities for observation at a greater
height,—we packed up that part of the apparatus, as
so much preparation, for our impending move to
"ALTA VISTA."   Such appeared to be the name
of the place I had now selected for our next station,
on the strength of the reconnoitring party's report.
The *locale* was on the Peak; seemed to be at least
10,500 feet in elevation; was the very last point to
which horses could ascend; and was more than 900
feet above the Estancia, the usual spot where travellers abandon their mules, and betake themselves to
climbing on foot.

Our interpreter had conscientiously been up to the
very top of the cone; but specimens of sulphur which
he brought back with him, and accounts which he
gave of the hot ground throwing off steam, showed
that it would not be worth while to overcome many
difficulties, to reach such a place with astronomical
intentions. Contenting ourselves therefore with the
height of " Alta Vista," we drew up plans for a more
regular and methodical arrangement of station, than
the hand to mouth affair on Guajara; and we proposed even the luxury of living in regular houses, by
dint of roofing in neighbouring walls. Remembering
too, how much we owed on our first night to the providential calmness of the air, and that we could not
be certain of the same good fortune again,—we determined to have some protection erected on the place,
previous to moving over with all the instruments.
Moreover one of our party having complained of not
having slept a wink all night, on account of his
neighbour talking so much in his sleep; and that
talk proving, on inquiry, to have been all about walls,
building them in this way or that way,—I immediately
put the accused at the head of a party for carrying out
again a work, which he had entered into so zealously
on the former occasion, as to have got it identified

with his very system, and made the regulator of his dreaming thoughts.

Meanwhile attempts were made to obtain some measures of the intensity and plane, of polarization of light from the blue sky; but they were no sooner commenced, than a cloud began to form over the station, and continued till night. The same thing occurred next day, and next, with an addition of descending wisps of rain seen beyond the Peak. The following day, a cloud again formed towards noon over our heads, and presently began the unusual sound of pattering drops of rain.

The air was excessively calm; all nature seemed to leave off every other occupation, and stand open-mouthed to catch, and make the most, of this shower. Our tent soon began to shrug its shoulders, as the wetted canvas contracted; and to tighten the guy-ropes, which from this cause, needed attention every few minutes. But otherwise, we delighted in the visitation; for the wooden boxes were being greatly improved by it, and gave promise of returning to their original dimensions.

The ultimate effect however was but very small and partial; for after nearly three hours of rain, excepting the side actually exposed and wetted, each box was as dry as ever; and the depression of dew point was never less than 22°. Our own stratum of air was in fact still dry and properly African; while the slight rain which had fallen to us, from the accident of our being under a certain cloud, had evidently been matured in an upper region of the atmosphere, with which we had nothing in common.

These "Southern" rains as they call them, are not unknown during summer in the vineyards on the coast; where their few and big, warm drops, just mark the passage of elevated S.W. clouds over the zenith of a place, and nothing more. The natives are not disappointed; for they seem to have learnt, according to true theory, that the moisture conveyed along at that height, is not intended for them; nor meant to fall on earth, until the whole stratum becomes a surface current; as it may do somewhere in the parallel of 40° of latitude, or beyond the range of their Trades.

Now and then the husbandmen do have heavy showers in summer; but these come from the N.E. cloud, at a height of 3000 to 5000 feet; and are not therefore felt in the upper parts of the mountain. One of these rainy days, no less unusual with the

Polar wind in Teneriffe, than in Scotland, seems to have occurred when Captain Basil Hall visited the Peak; and the appearances which he describes of festoons and hangings of fringe from the clouds below him, had not anything to do with the general appearance of clouds seen from above; but were simply owing to the circumstance, of there being rain that day, between the level of 4000 feet and the sea.

The Trade-wind is undoubtedly a poor one for bringing water; but its position in Teneriffe during summer, is favourable for making it deposit any, that there may be present. Hence these occasional wet days, over the lowlands in July and August. Hence too our friends at Santa Cruz and Laguna, used in those months to write dismal letters to us, sympathising with the frightful weather which they thought we must be encountering on the mountain; at the very time, that we, above the level of all the disturbance, were in reality enjoying ourselves most, under fine skies, and sunshine surpassingly bright.

Letters came also during this week from the Edinburgh Observatory, describing how a series of nothing but cloudy days, S.W. wind, and rain, had prevailed there ever since we had left.

These were no doubt, the produce of that upper Equatorial current; which, looking upwards from Guajara, we had long seen day after day, ever sailing steadily, but nearly transparently, on, far above the mountain tops.

CHAPTER VIII.

END OF GUAJARA.

AFTER several nights of unusual cloud, the evening of August 15th displayed a clear blue vault of sky, and the full moon starting from behind the purple summits of Grand Canary, commenced to establish almost a second day. The cloud referred to as unusual, was of course composed of the S.W. cirrocumuli, floating high in the air; for the lower N.E. cumuloni were never disturbed, from month's end to month's end, from their sluggish position below.

That there is a preponderance of clear weather on the night of full moon, and that such clearness is due to a dissolving of vaporous clouds, by heat reflected from the lunar orb, are ideas started by Sir John Herschel. While if he and other experimenters, have been unable with their most sensitive instruments, to detect any symptom of that heat, the want of success, he has suggested, may be due to the caloric being expended in upper parts of the atmosphere, in this very dissolution of vapour. And owing to its original low temperature, such lunar heat would have great difficulty in passing through, even a transparent medium.

Some confirmation of these views was at once found on this occasion; for the upper clouds had vanished before our eyes, under the presence of a full moon; and if the lower ones had not also disappeared, that might be considered only as a consequence of the higher strata, having exhausted the full potential energy of heat thrown upon them. A person below those N.E. clouds then, even had he caught a glimpse of the moon through an opening, should not have expected, even with the most delicate apparatus, to be able to detect any traces of heat. But raised as we were on Guajara, 4000 feet above the mist, our chance of getting a positive result was evidently far better. Moreover, the only instance known of any real success attending experiments on lunar heat, is that of M. Melloni; when employing not only his invaluable improvement on the thermometer, the thermomultiplier, but taking his stand on the cone of Vesuvius, at a height of 3000 feet above the sea level.

Melloni's experiment hardly seems to have received

all the attention which it deserves, in this country.
Whether his plans for concentrating the lunar rays,
in the first instance, were thought exceptional; or,
whether simply his success, in a path where every
one else had uniformly failed, was considered to need
confirmation, before being finally accepted as a fact
in nature,—I do not know.   But I prepared on this
evening, to try, under better circumstances than his
as to height, and without any condensing apparatus,
—whether the heating effect of the moon's rays
might be sensible to an excellent thermomultiplier,
kindly lent me for this purpose, by the well-known
electrician J. P. Gassiot, Esq.

As the moon gradually rose higher and higher,
some observations of its optical spectrum were made ;
chiefly remarkable for the quantity of blood-red light
thereby proved to exist, in its innocent-looking blue,
or rather violet-coloured rays.   At last, about 11
o'clock P.M., the heat experiments were commenced.
The moon was in unfortunately low declination, so
as to have a meridian altitude of only 42°; but all
other circumstances were eminently favourable.   The
air was pure, and perfectly calm, every one but myself
had long gone to bed, the fires had been put out four
hours previously, and their sites were a long way off,

with stone walls between. So the apparatus being mounted on a small pier inside the telescope enclosure, with a range of nearly 20 feet clear in front of it, with no artificial lights about, (for I had found it possible to read off the graduation, and write down the figures by moonlight) ; and with no other active source of heat in the neighbourhood, than my own body, and that swathed abundantly in non-conducting flannel, and kept well away from the thermotic pile,—I tried a preliminary experiment.

Holding my naked hand in front of the peculiar voltaic arrangement, and at a distance of three feet, there was an instant move of the magnetic-needle through 7°; and then bringing it within three inches, there was so large a deviation, that I had to wait a long time, before the needle had recovered from the disturbance.

How calm, how perfectly dead calm was the air all that time; not a breath could be felt; not a sound heard; there was the silence, and stillness of death. This degree of silence felt inappropriate on a high mountain; for on such, there is in general, so ceaseless a murmur of hundreds of torrents far and near, working their way downwards continually; and never for a second leaving off their bubbling, splashing, struggling onward;—when powerful, even urging on the stones in their beds with a perpetual low, grinding, rumbling noise. But on Teneriffe there was nothing of this sort; the absolute aridity of air and ground, had denied the existence of a single stream. A faint tinkle tinkle now and then, from a stray goat, was the only sound to be heard during this anxious period; and though the creature was far off, one could distinguish whenever it stopped to browse on some solitary *retama* bush, and then when it trotted off to find another.

The needle came at length to rest; so, quickly turning the voltaic cone, which had been directed to clear sky, on the moon,—I anxiously watched the result,—the needle scarcely moved. Lunar radiation was small indeed then, and I girded up my loins to try special methods of observing, suitable to such a case. The plan decided on at last, was to take a large number of readings, at stated short intervals, combined with variations in the direction of the cone. Having at length obtained about two hundred such, in the course of an hour and a half, I was extremely well pleased to find, that the mean of the numbers indicated an undoubted heat effect, of about a third of a degree.

Had the recording instrument been a Fahrenheit's thermometer, the whole operation would have been concluded. But as a thermomultiplier's degrees may be almost anything, I immediately placed a candle on a stool, 15 feet in front of the pile; and observing it on exactly the same principle as the moon,—there was given a heat effect of nearly one degree.

With this result, there need be no wonder at the failures of former observers in England, near the level of the sea, and before the day of the thermomultiplier, —to obtain any instrumental evidence of the moon's radiation; for here, on so peculiarly favourable a position, with the luminary shining away quite blindingly, the heat was only one-third that of an ordinary candle, at the distance stated above.

After having obtained a nearly similar result by about ninety observations on the following night; the packing up for our intended move, proceeded so strenuously, that on the morning of August the 19th, when a visitor was announced, there was little of a scientific nature to be seen. This we much regretted, as our new acquaintance, Don Martin Roderiguez, proved to be a most intelligent and well-

informed man; and in what a fine old Spanish style
he came up. First there arrived two men with a vast
supply of goats' milk as a present for the sailors,
ourselves, and every one on the place; then there
followed attendants with loaves of splendid home-
made bread; some, white and light as that of Paris,
and some, of a richer description, saffron in colour,
and with a crust, that happily did not become casehardened, in the dry atmosphere above the clouds.

Then too, when we invited the Don to join us
at breakfast, he produced his quota in the shape of a
dish of partridges,—flavoured with miniature cloves
of choice garlic,—and home-made cakes, which were
a compound of hardly anything but almonds and
honey. He had hoped to have caught us a rabbit
or two, but had lost his ferret in the attempt that
morning, not very far from the tents. Without
such addition however, there was more than enough
for a feast all day long; and when, over and above so
much, his herdsmen arrived in the evening with more
of the goats' milk, as luscious and as thick almost
as cream, we had to empty out the water stowed in
tins and bottles for our intended journey to Alta
Vista next day, in order to profit by these unexpected
supplies of the richer fluid.

MASSES OF LAVA SLAG AT ALTA VISTA

p. 260

Printed by A. J. McIntosh under the superintendence of James Glaisher, Esqr F.R.S.
and published by Lovell Reeve

With the larger instruments packed up, there was more time than before, for attending to peculiarities of twilight, the zodiacal light and similar eye phenomena. The zodiacal light, ever an interesting object, was particularly so just now; not only because this was the first good opportunity that I had had of supplementing my observations in the southern hemisphere; but on account of certain new ideas recently published, and not a little hostile to the dynamic theory of heat. They tended indeed, to deprive that mathematical conception of one of its cosmical triumphs, viz., an explanation of the true nature and origin of solar heat. This solution of a problem, older than Zoroaster, had seemed to receive decided confirmation, when theorists could point to the zodiacal light, under an heliacal hypothesis, and hold it up as a proof of numerous meteoric stones falling, and about to fall, into the sun; thereby having their dynamic energy, resolved into its equivalent of heat and light.

From observations lately made however on the American expedition to Japan, a conclusion has been boldly published, that the zodiacal light encircles, not the sun, but the earth; which should in consequence have something of a Saturnian look, as seen from

other members of the solar system. No sooner was
this new hypothesis put into print, than it was endorsed by several distinguished names; and the Abbe,
conducting the Journal " Cosmos," stated "that a
notion of the same sort had long been caressed by
himself."

If the observations were good and unique,
of course their result must be received at once,
despite any beautiful theory, built up on an abstract
idea. But there are many older observations existing, by no means bad, and pointing to an opposite
conclusion; while the phenomenon is by no means
easy to measure with certainty. Indeed a more
southern position, and a clearer climate than England are so necessary to unexceptionable work, that
home astronomers have generally deferred entirely
to the accounts of travellers. Now our position on
Guajara being more than usually favourable, even for
travellers, in latitude and elevation,—what did we
see,—asked several of our friends?

In the evening, after twilight had ceased, we saw
the strange glade of light, very fairly, though the
angle of the ecliptic was such, as to allow the axis
an angle of only 31° to the horizon; and hence, but
for the purity and rarity of air at 8900 feet of eleva-

tion, nothing would have been made out. As it was, we observed instrumentally, night after night, all the several elements of length, breadth at horizon, and place, or R.A., and D., of apex.

Gradually the inclined tongue set in the West; and for several hours before and after midnight, we were left with only stars and milky way to admire. But at two o'clock, something began to appear in the East; and at four o'clock, there was a gorgeous zodiacal light at an angle of 75° with the horizon, and with the length of 63°. So bright was it towards the base, that it produced a weak reflected glow in the West; and we could occasionally fancy a tail, of the faintest conceivable light, extending nearly to the zenith. Nevertheless there was no doubt of the lenticular form of all the chief mass of light; and the place of its apex, as measured, was always consistent enough.

In this there was confirmation rather of the old, than new, ideas; but one evening (August 18th), looking towards the Eastern horizon, where the moon was presently to rise, there was a cone of light visible; and so very "zodiacal" did it seem, that we had almost jumped up, and said, that the question was now so far settled against the old school. The appearances followed in this order; at 17h. 5m. Sid. time, the moon's twilight was perceptible as a low flat elliptical arch of faint light; at 17h. 12m. it had manifestly grown obtusely pyramidal, or somewhat pointed above; at 17h. 15m. the point had extended itself into a cone, all of the faintest light, 30° high and some 12° broad; and at 17h. 20m. the moon rose.

What, however, all this time was the angular position of the axis of the cone? On being measured, it invariably proved 90° to the horizon. I was rather surprised at this, for the general appearance to the eye was so similar, on the whole, to the zodiacal light, that I had felt almost certain of an inclination; as probably any one else would have done from the mere impression of his senses. On similar judgment, every untutored person, believes that he sees the sun larger at rising or setting, than when high in the sky; and yet the case is really the reverse of this, as angular measure instantly demonstrates. In the same way, each of my measures in the present case, proved the axis of such lenticular cone of lunar light, to be standing vertically on the horizon.

Now this circumstance, combined with the further one, that the angle of the ecliptic in that part of the

sky, was only 38°, not only showed that the appearance
was not caused by that suspected cosmical body, which
with solar illumination, causes the legitimate zodiacal
light,—but that it was some local affair, depending
on atmospheric refraction and lunar dawn.

Seeing also, that the cone became visible only *after*
the lunar twilight had manifested itself strongly, and
immediately before, the moon rose,—the solar analogy
which it bears, is not to the zodiacal light that
vanishes with the very first dawn of morning,—but,
with the pyramid of rose-pink light, which subsequently surmounts the more level arrangement of
orange and yellow in lower parts of the sky, when
the sun is about to appear. This feature is well
known, and very dear, to painters; but has little interest for astronomers. The manner, and relative
times of showing, are precisely alike with these two
phenomena, the rose-pink pyramid and the white cone;
while if one of them be so richly coloured, and the
other not at all,—it is quite agreeable to the difference
known to exist, between solar and lunar rainbows.

The concluded lunar zodiacal light then, was not
only erroneous in itself; but, when duly considered
with regard to its real solar analogue,—was a sufficient proof that the human eye, would have been

unable to appreciate anything so inconceivably faint,
as must be a lunar analogue of the solar zodiacal light;
supposing the latter for a moment to depend, on the
sun illuminating a nebulous ring, encircling, not
himself, but the earth.

A more startling feature still, however, than any
lunar phenomenon, had been mentioned by the pro-
pounder of the new theory in favour of his doctrine;
viz., that as he had sailed in the tropics, from one
side of the equator to the other,—the zodiacal light
had accompanied him, changing its place amongst the
stars.   This is in itself a mighty and a conclusive
fact, regarding what the observer himself saw.   But
when I found my observations for the place of the
light on Teneriffe, agree with those made at the Cape
of Good Hope on the other side of the equator; and
with others procured in India, France, and England,
by a variety of astronomers during the last century,—
I could only come to the conclusion, that the zodiacal
light of observers in general, was not that of the
third volume of the "Expedition for opening up
Japan."

After our friend the Don, hospitable though not in
his own house, and giving an excellent idea of a

Spanish welcome, had left us on the evening of August the 19th, there was an immense boiling of milk going on at several different fires, in order to enable the fluid to keep during the next day's journey, under a hot sun. There were gallons of milk, and only two very small portable saucepans and a tin coffee-pot, to boil it in; so the stewing went on far into the night.

Gusty winds, from the N.E., arose; the fires flared, and crackled, and shot up clouds of sparks; but most of our tents were already taken down, packed up, roped tight, and stowed away within the quondam telescope enclosure. Freely, therefore, and without stint at their mountain fire-place, amongst rude masses of trachyte lava, with the white moon shining on them from above, and reflected broadly from the silvery sea of cloud below, the sailors heaped on *retama* wood. The flickering flames leapt up on high, revealing, in their lurid light, more and more swarthy faces every moment; for the men from Chasna were arriving with their mules, in preparation for the morning's move.

Not a very pleasant set were they, these Chasna men. Consumers of odious tobacco, and bad garlic, in overpowering abundance; they tossed our packages

about somewhat impudently, and then went back to
the fire denouncing their weight, and declaring the
impossibility of such burdens being carried on the
backs of any mules. What with the boisterous wind,
and the growling men, no one slept particularly well
that night on the mountain.

Next morning at an early hour, after a refreshment
of the Don's rich goats' milk, our last tent was
struck, and the Chasna men were called on to load up.
One or two did, but others refused, without more pay;
and then they all began disputing with each other
and the interpreter.

After a couple of hours of violent vociferation, and
vehement action that must have made their arms ache
fearfully, and after some of them had gone back,—an
arrangement, not very wisely on our part, was concluded with these villagers,—the loading was re-
sumed; and by 9 o'clock A.M., some on horses, and
some on baggage mules with wooden saddles, we all
rode off, from the top of the "wind-loved" Guajara.

PART III.

ON THE CRATER OF ERUPTION.

CHAPTER I.

SCALING THE CENTRAL CONE.

WITH our long train of mules defiling down the
pass, we soon reached the floor of the great
crater, whose central cone we were to climb before
night. The direct distance to be travelled, was but
four miles, and the general inclination nothing
important, but the roughness was verily incon-
ceivable. One of our reconnoitring parties a few
days previously, had tried a straight cut across this
lava-covered plain, to save themselves the trouble of
going round towards the east, to a smoother region;
but after awhile they became entangled amid such
terrible stones, that they had to take all the baggage
off their mules, and carry everything on their own
shoulders.

As we stood on the Canada, the pumice beach of a
once fiery sea, its frontal wave of lava, rose between
us and the Peak, as a long ridge of rocks piled one
over the other, at the steepest angle at which they

Q

could avoid falling over. Looking westward, the
lava rolled up to the crater wall, formed there by
Guajara, whose cliffs and avalanches of broken stones,
we loved to trace from the summit, down to the base;
where sudden pinnacles and spires shot up, with a
strong family resemblance to the " Lunar Rocks."
Some appearances of this phenomenon were seen also
in the eastern direction, whither we now pursued our
way; gazing upwards in admiration at the range of
mural precipices far above our heads.

The lower parts of these ramparts of the great
elevation-crater, were generally but steep slopes of
stony debris; yet being bared in some places, the
construction of the mass was illustrated in fine
sections of horizontal strata of green-stone, and
tufas; some of them hundreds of feet in thickness,
and others thinning away, like ribbons, to a few
inches. These latter looked often at a distance, so
like the veins in the sides of an alluvial sand pit,
that one was inclined to examine them for the much-
desired fossils of Teneriffe.

Von Buch considers the great crater, says one, who
is no friend to the author he quotes, to have been
formed under the sea, although no fossils have ever
been discovered upon it. This is a cruel dig, and is

it deserved? Why not, adds the critic, what is the proof of formation at the bottom of the sea, but finding marine organisms in the rock; have not such been found on Monte Somma; and has not Teneriffe been compared to Somma? Yes truly; yet all the analogous parts of each have not been examined. Fossils are by no means found over the whole of Somma, but only on those parts which are represented in Teneriffe by the southern foot of Guajara, a region not yet investigated by good geologists.

The stereotyped route for ascending the Peak, is from Orotava; and unfortunately lies over nothing but sub-aerial lavas and pumice; somewhat as if one were to ascend Vesuvius from Torre del Greco, and declare that all is barren. Barren in good truth, and without organic remains, should all such slopes be found, when we walk over streams upon streams of material, that have flowed forth at the temperature of melted iron, from a caldron seething for ages with fire and liquid rock, and incandescent acid gases.

Wherever we examined the beds, forming the submarine crater walls, crystallizing effects of heat were predominant In one case indeed, a stratum of greenstone exhibited the once action of water, in having cracked and rent into little pieces, only two or three

tenths of an inch in the side; giving it the appearance
of hot glass dipped suddenly into a cold fluid; or that
stream of lava, which Dana describes, flowing into the
sea from Kilauea, and crackling into fragments as it
went. Yet even in this case, the greenstone or basaltic
lava, had not only its own ordinary indications of being
a fire-formed rock, but by the numerous long crystals
of glassy felspar which it contained, torn and bent as
they were, seemed to prove that it was a melting of
some previous mineral, whose refractory felspar crystals had alone resisted the second application of heat.

Such strata of the mountainous walls on our right,
were objects of unceasing interest in their varieties of
colour, and their accidents of faults and dykes.
These last had sometimes altered the appearance of
parts through which they had pushed their way;
but were more noteworthy, when considered in connection with the well-known lunar striæ of Tycho,
extending unbroken over hill and dale,—from their
having always conformed to the surface of the
rock, whatever its inclination; and not to the law of
fluids seeking a level. In some places, the dykes
certainly stood up above the matrix; but plainly, from
a modern wearing down of the latter.

On our left all this time, were the frontal ridges of
sub-aerial lava streams; sometimes pressing so close

up to the walls, that our horses had great difficulty
in stepping through heaps of tumbled stones.
These stones, massive as they might be, were but the
foam-flecks of that fire-borne flood, which surging
and resurging like the ocean below, had in ages long
since, covered all the area around the central Peak,
with colossal ruins.

Occasionally amongst the brown blocks was found
one of charming sea-green tint, smooth, dense, and
in a mighty lump.   This had tumbled from the cliffs
above; and in its nature, appears to have been a
product of the ancient submarine volcano only.
Similarly one or two huge portions of a porous lava,
in colour a rich lakey Indian-red, and as pure and
true in tint, as if specially prepared by an artist's
manufacturer.   I broke off portions as a great prize,
and packed them away in a knapsack, together with
stones of cobalt-blue, lilac, purple, green, orange,
yellow-ochre, and Naples-yellow, with some view to
converting them into pigments.   How inexpressibly
varied and brilliant were the colours then, bathed in
the light of a vertical sun, undimmed by impurities of the lower atmosphere; how dull and monotonous do they appear now, under the grey of a
Scottish sky.

As we gradually made our way towards the eastern

side of the crater, the long ridge shape of the Peak, seen over hummocky heaps of yellow lava, slowly shortened; the abutment of Montaña Blanco, with its deep red streams and their viscous wrinkles, was projected on the central cone; and then, when at last, it began to stand out to the south of the Peak, —there was seen in the midst of rearing bergs of red and brown lava, of masses of tossed and tumbled rocks, of peaks sharper than the *aiguilles* of Mont Blanc,—a fine parasitic crater of intense hue, and one would think, of immense potential energy. The tops of other such craters appeared here and there in the distance; while behind and above all, rose the grand Peak, seamed with its blue-black torrents, and show-ing clearly the dimensions of its once active crater of Rambleta, at the height of 11,700 feet; while imme-diately in front lay a glittering sea of pumice, dotted with grotesque groups of reddened slag.

Allowing the rest of the party to proceed, my wife and I stopped to sketch this rare volcanic scene, though the sun had become furiously hot. The place however was not without its *agrémens*, for fine speci-mens of *retama* were growing round about; few and far between, yet princely in their almost globular form, as compared with the flat and stunted bushes on Guajara. Presently too a herd of goats came past, led by a glorious old animal of the Guanche breed; Vandyke brown in colour; with long twisted horns, a venerable beard, and hair lengthening almost into a lion's mane.

A few steps beyond this spot, was a large flat-surfaced rock, right in the pathway, and so splendidly grooved and scratched, as to be rather too like the marks of a clogged waggon wheel; but such a machine had certainly never been there, so we secured a portion. A little further, and the "coup d'œil" was twice as fine as before; the small crater had opened out its mouth, and now showed a most inordinate swallow. But we could not stop to draw again, for the rest of our party were out of sight, and this was exactly the point where we had to enter the lava-strewed plain, and strike across to gain the Peak.

We entered, and soon got surrounded by jagged and impassable rocks; our attendants, guides though they professed to be, were at fault. They ran here and there to look for a path; then led the horses through hollows, where rough rasping points of stone peered everywhere above the thin sandy covering;

and then they climbed up to the highest barrier in the neighbourhood, and hallooed with all their might, but without avail, to those who had gone on before.

The sun was now on the meridian, and the altitude of our position being upwards of 7000 feet, the radiation was something scorching; and both horses and men were wearing out in scrambling attempts to recover the path. At length a trace of footsteps was struck; and away we went on a curving line, in and out amongst rocks of every conceivable shape, always, however, treading on a sprinkling of fine pumice. In this course we approached the small crater,—from its colour, one was inclined to fancy it composed of good garden mould; but as we passed over its southern foot, we found nothing but the hardest of clinkery cinders. Away, and still away we kept urging our horses on the "spoor;" the reefs of rocks and rugged piles of stones became more complicated, but happily their crevices and gaps were always nicely closed by pumice-stone dust. Everywhere was it sifted down, yet so moderately, as to leave no more than just room enough, to enable a clever horse to walk along without endangering his legs; his feet at the same time, leaving a mark, legible to a countryman skilled in tracking.

Before long, overtaking the troop,—horses, mules, and men were found enjoying a sort of standing rest in the open sunlight, on the top of a hillock, with red crags rising through its glaring pumice-stone soil.

From this point we overlooked the whole south-eastern floor; stream on stream of lava, rolling its ridges of stones over its fellows; here and there a small eruption crater of dark brown; and then another larger one of the red period, of firmer rock, but still so rent and cleft, that the caldron's circle looked like some mighty Druid's temple. But in that case, where the oak-trees or the grass that Druids loved so well?

Here, all around, excepting an occasional shrivelled-looking *retama*,—a plant that grows naturally without apparent leaves, and with merely a bristling collection of glaucous twigs,—there was to be seen nought, but hot red rock, and thirsty yellow pumice; while the scorching sun over head, and the blue un-varied sky, with its uniform saddening tint of arid light, seemed to condemn everything, far and near, to barrenness and desolation for ever.

In such a wilderness, one fresh from English meadows, might exclaim, oh! for " the shadow of a great rock in a weary land." The phrase expresses much that we felt at the moment; and was far more suitable there, than to the case of a recent Himalayan traveller, who quotes it while spreading his blanket at night, under the warm lee of a large block, in one of the snowy passes of those mountains; but the figure is only fully applicable, in an Asian or African desert. In glowing "Karroo" plains for instance, where one travels day after day over a nearly level surface, with a horizon like the sea; and at last may meet with some great protruding rock, a mass it may be of quartz, such as appears occasionally in the Hardeveld country, rising through salt land, and uninhabited; and forming the one lone beacon, the only thing with shade, in all the region round.

In our contracted, and walled-in, desert of the great crater, a more suitable illustration seemed to be, " the line of confusion, and the stones of emptiness." Confusion worse confounded than that of the jostling lava streams, there could hardly be; and if the stones of emptiness be not pumice, which is so full of vesicles as to be lighter than wood,—the case is one of the many, wherein a Bible simile is found to have not a few striking natural applications, as well as that particular social one, it was specially intended to represent. Bishop Lowth indeed says in his Commentary, that the stones of emptiness refer to the plummet of the builder; but plummets being usually made of lead or other heavy metal, this explanation is not immediately convincing.

The sun was now inclining westward; and on our venturing to look up to the pierced sugar-loaf cone at the summit of the Peak, there was volcanic activity, let loose once again; for puffs and jets and rings of smoke or steam followed each other in quick succession; playing away as actively as if from the discharge-pipes of a number of high-pressure steam-engines.

At this sight, eagerly pushing on again, we emerged from the chaos of lava barriers, and found a pathway ascending the side of Montaña Blanco; a strange creation. Upwards of 9000 feet in total elevation, this mountain had yet the gently rounded outline of any little mouldering hill of silt or sand. The soil itself was curious; Naples-yellow in tint, and composed of nothing but lumps of pumice stone, about the size of a walnut, and as light as water; how then could it resist being swept down by the winter's rain?

Distantly marking the surface of the corn-coloured
Montaña, were seen several black rocks, whose angles
shone and blazed in the sun-like heliotropes. Now,
for obsidian *in situ* at last! But it was no such thing.
The first of them we passed, was more than ten feet
cube, with some of its surfaces, those that reflected the
sun into our eyes from a distance, eminently smooth,
as if with crystalline cleavage; yet the structure of
the mass was decidedly stony, not vitreous. Black as
coal, though relieved here and there by small white
crystals of felspar, these huge boulders looked
strangely "travelled" on the yellow soil; and some
of them were so far up Montaña Blanco, where its
slopes opposed those of the Peak, that we could hardly
at the time think of them as pebbles, fallen from the
lava torrents of Rambleta.

These we were now approaching; huge embankments they looked, nicely bevelled off at the sides and
ends, as we had already seen from Guajara. While
as if to confirm the chronological and thermotic
theory there arrived at, a large stream had stopped
short on one of the steepest parts of the Peak, where
the slope was by measure 29°; or more than double
the angle of the exterior walls of the great crater;
and at least ten times the inclination of parts of its

floor, where the earlier yellow lava had run freely
enough.

Winding along next in a sort of valley, that led
upwards, between Montaña Blanco with its round
yellow head on the left, and on the right the ashy
flanks of the Peak, and the lower ends of black lava
cataracts close above us,—we at last came to a part,
where they crossed the road, closed up the head of
the gulley, and barred all further progress in that
direction.

Some of the muleteers who had already arrived at
this spot, were anxiously looking upwards ; and there,
between two of the dark ridges, rose every now and
then, clouds of smoke or dust. Which of the two,
we could not tell ; for with the sun now directly in
our eyes, we could hardly distinguish the true form
of the ground ; while the vapour was glorified, as it
were, over and above measure, by the strong solar illumination drenching, and transfusing, it with light.

On reaching this point ourselves, we found a narrow
strip of the original surface of cinder hill, stretching up as far as we could see, and closely bounded on
either side by black torrents of lava. Now began
the work, for we had fairly arrived at the cone or
crater of eruption, with its average slope of 28° : and

though the path zigzagged from side to side, and the men showered imprecations and blows on their mules, and took half a turn in their tails as well, there was an obligatory stopping to rest and blow, every few minutes.

The scene all round about, did now indeed appear volcanic; and all that we had met with before, was but ideal or theatrical, tricked out with so many bright colours, and decorated to some extent with vegetable life. Here the bushes were gone, we walked up a surface of mere ashes of pumice, and on either side were tumultuous cascades of stones as black as night. They had been poured out probably before the age of man; but were still so naked, so purely black, so exactly as they had tumbled about at that very period, that one could have fancied them an exudation of the mountain only yesterday.

Continually did natural cleft surfaces of some of these blocks, glint in the sunshine like looking-glass; and as often I left my horse on the ash slope, and clambered over piles of loose clinkers, hammer in hand; but always to find that we had not yet arrived at the parentage of obsidian; the material before us was still the granular lava.

At 4 o'clock we had reached "Estancia de los Ingleses," at the altitude of 9700 feet, the usual sleeping-place of routine travellers; who camp out here on their first night from Orotava, and ascend the remainder of the Peak next morning on foot.

At this point there is something of a shoulder in the slope of pumice, and on it some huge blocks of lava have fallen from the Northern ridge. Could some of them be hollowed out, they would form very excellent sleeping apartments; but as it is, such protection from the wind as can be found on their lee-side, is all their claim to be considered an "Estancia." Shelter from the cold heaven above, they certainly afford none, and it may be doubted whether the heat which is impressed on them by the sun, during day, is radiated out, as some travellers say, to a very sensible extent at night. Be that as it may, the greater part of men trust rather to a fire of *retama* wood, some few specimens of which plant still maintain themselves in the loose cindery soil, up to the elevation of 10,000 feet. But on the lateral hills of black stones forming the streams of lava—there is absolutely nothing.

After leaving Estancia, the ascent became sensibly steeper; the pumice stone too, looked newer than what we had been accustomed to on Guajara, and exhibited a filamentous glassy structure.

The ridges of unearthly black stones on either side, as we went on, gradually approximated; and now that the sun had set,—looked blacker than ever. At length between 5 and 6 o'clock, P.M., they had closed in upon us in front as well; and then we had reached "Alta Vista," the *ne plus ultra* of beasts of burden, at an altitude of 10,700 feet.

CHAPTER II.

ON the 21st of August, we rose early, excited alike by the novelties of the neighbourhood, and by the occurrence of another term-day. Our interpreter and his men had done their work famously in the preceding week, having erected altogether something like sixty yards of wall, four feet thick and six feet high; and so arranged, as to form a telescope enclosure in the centre, with cells or small rooms round about. One of these, intended subsequently for optical experiments, was soon furnished with half a roof over its southern end; and there we established the meteorological instruments.

The barometer was at 20·5 inches; the thermometer at 6 A.M., was 49°; and at noon, its maximum, was 65°. The air was thus found very sensibly cooler here, than on Guajara; while the dryness was rather less, the depression of the dew point varying from 18° at night, to 41° in the middle of the day. The

R

power of radiation, however, appeared to be increased, and the inclination therefore, in the temperature, to mount up to a maximum with noon, became its marked feature.

The earlier part of the solar course, or from the eastern horizon upwards, could be well commanded at Alta Vista; but at 3h. 45m. P.M., the sun was lost behind the western lava ridges of the Peak; while other descending streams, on the immediate N. and S. of our station, cut off the lower part of the sky in those directions also.

Our look-out was thus limited; but grand and imposing nevertheless. Eastward, gazing down the slope of Estancia de los Ingleses and Montaña Blanco, we saw the red and brown lava confusion of the "floor," with little cup-shaped craters here and there; and beyond that, the regularly stratified walls of the great crater, with their curving bay-like recesses, appearing at this distance as mere conchoidal fractures in a hand-specimen. Then further still were seen purple high lands forming the backbone of the island, and stretching away as far as Laguna—perpetually surrounded with clouds—and even to the Peaks of Anaga beyond.

At intervals along this whole line, were seen parasitic,

or rather small eruption, craters. It was one of these which opened in 1705 above the town of Guimar, and sent forth an immense stream of lava, that rolling down towards the town, divided apart just above it. In this way, enclosing the unfortunate city between two burning rivers, the ejection flowed onwards, towards the sea; and had well nigh cut off from the inhabitants every chance of escape, except by running the gauntlet either of fire or water. While this was going on eleven miles to the east of the Peak, a similar eruption broke forth seven miles to the N.W.; and there, without scruple, went right into the town of Garachico, and filled up its beautiful bay.

When we thought of this sympathy on either side of the great mound, and saw the mouths of " rapilli" craters, wherever any part of the country was at all visible, we could but think how this entire island must be undermined by the volcanic forces, whose crowning vent was now close above our heads. We were even startled to find how close; for by stepping out only a few feet in front of our walls, we could see, appearing above the black lava ridges, a small portion of the sugar-loaf cone, the culminating point of the whole mountain. Intensely bright did it look, and so yellow, one thought it but a few feet off; while even

and anon, white jets of steam shot up in the rarefied
air, like miniature fireworks.

In connection with the meteorological observations,
we had rather to look downward, than upward. Above,
there was only flat blue sky; but below, a varied sea
of clouds spread out over the surface of the ocean; as
level, and almost as broad. From our greater height
now, we could still better judge of the vast extent of
the misty stratum, than at Guajara; and although
Palma, Gomera and Hiero were no longer in view,
being hid behind the Peak,—the 7000 foot mountains
of Grand Canary were fully commanded; and made
an excellent measuring rod, for ascertaining the height
of the sheet of vapour. It proved to be very uniformly
about 4500 feet.

This sea-cloud too, we could now very clearly see,
did not abut on any part of the higher mountains of
Teneriffe; which had on every side of them, a certain
thin stratum, of their own as it were, at a level of
3000 feet. In the early morning, and late in the
evening, this land cloud, duly preserving its level,
stretched up to, but did not pass, the vertical confines
of the sea cloud. Then it seemed with its smoother
surface and thinner structure, to represent, in mist,

the "ice-foot" of the Polar regions; that edge of ice
adhering tenaciously to the sea cliffs, as described by
Dr. Kane; and offering on its smooth and stationary
surface, a safe asylum for his poor persecuted little
party and their battered boats. There they always
found a refuge, whenever the straits were filled with
bergs and broken floes, that went sailing and crushing past in grand confusion, and in an endless irresistible stream; just as indeed, in their way, the
ocean cumuloni were now hurrying tumultuously past
the vapour-foot of Teneriffe.

Well could we discern these strange aerial doings,
but better nevertheless was our station for scientific
supervision, than for pictorial effects of this world of
cloud-land. For, standing as we were over the
centre of an eight mile broad crater, whose walls
dipping at an angle of only 12°, were at their summits full 4000 feet above the mist stratum,—the
very nearest of the clouds were six to seven miles off.
So far, in fact, were they removed into the distance
of the landscape, and such magnificent solid ramparts
and broad plains did we see between us and them,
that in spite of what has been written by some
ascenders of the Peak, we could not conjure up their
feeling of having a second flood before us; or of our

being the last lone men on the topmost pinnacle of some high mountain, of which all but a few remaining yards, have been submerged under the sea of vapour.

On such an ideal rock, no doubt the approach of this cloud-deluge is grand in the extreme. You have observed it perhaps, from its first advance with a few scattered rollers, which fling themselves against the precipitous rocks; dash up like spray; and get partly dissolved by the warmth of the soil. Then more and more of the masses pour in; they cover all the lower country, and the habitations of men; and as they gradually rise in elevation, you see portion after portion of the surrounding hills ruthlessly swallowed up. At last perhaps, the sole remaining fragment of solid earth in sight, besides that you stand on, is some similar crag, that once formed the summit of a towering mountain. Everywhere else, is nought but the cold whiteness of rollers of mist. Should the sun be then in the act of setting, leaving you to battle in darkness against the growing inundation, on which, remember, you cannot float away,—the mental effect is such as is worth going far to experience.

There are such stations in the real, as well as the ideal; and the rocky "bergen," 3 to 4000 feet high,

Photo-Stereograph II.

SPECIMEN OF THE MALPAYS OF BLACK LAVA NEAR ALTA VISTA.

p. 246. 297. 363.

Printed by A. J. Melhuish under the superintendence of James Glaisher, Esq. F.R.S.
and published by Lovell Reeve.

on the west coast of the Cape colony, separated from each other by flat sandy plains, occasionally afford this spectacle, to one who dwells long on their summits. But the Peak of Teneriffe can never do so during its investment by the N.E. Trade-wind, its normal condition invariably through the summer, or most usual time of visitation by travellers; for it is too high, and raised on too broad a base.

Yet those who ascend the Peak for once in their lives, and have never seen the upper·surface of the clouds before, write many rhapsodies. No one perhaps has discoursed more beautifully on the subject, than a celebrated traveller; who, trusting over-much to memory of a single visit paid long ago, brings the sea of clouds up from 4000 feet, to an altitude of 11,700; i. e. to the base of the sugar-loaf, on the summit of which he was perched in the very middle of summer. By such disregard of measure, he presumes almost as great an infraction of nature's laws, as if he had made the water itself to rise and overflow the land. Had the summer cloud-level been really at that height, no blue sky would have been visible to us on Alta Vista; but in place of such a calamity to an astronomical expedition, we found heavens always clear; the sun, in radiation, outdid its Guajara ex-

ploits; and the old horned moon shone on us glitter-
ingly through bright daylight, until she set behind the
lava ridges of the Peak.

What a mystery did those ridges appear; no solid
rock anywhere, no soft earth, nothing but black
stones wherever we looked. The whole mountain
above us seemed to be composed of nothing but these
black stones, piled on black stones, and these on
black stones again. So loose was the piling, that
under the crenellated edge of one of the sky lines,
there were frequent glimpses of daylight; and at
night, after a star had apparently set, it reappeared
again; or would even advance far into the solid
substance of the mountain, before final disappearance; like those much-disputed cases of red stars
occulted by the moon.

Thanks to the dividing of a large stream of the
black stones just above our station, and the two parts
flowing down on either side of, instead of over, the
Alta Vista with its soil of comminuted pumice,—we
had been enabled to enter thus far into the sombre
wilderness, and contemplate from our pile of baggage
that further region, where neither horse nor mule may
venture. (*See Photo-stereograph, No.* 10.)

The sides of these ridges were at an extremity of
steepness for their material; so that a man going
up certain parts of them, brought down showers of
rubbish. There was no difficulty however, in his pick-
ing out such large stepping-stones, that his weight
made no difference in their stability. Some very
large blocks had fallen upon the Alta Vista platform,
happily before our walls were built; probably long,
long ago; and now—when we looked downwards to
Montaña Blanco, and saw the lower ends of the
black streams, less by degrees and beautifully fine in
the distance,—those immense and far-between "travelled" blocks, that had so excited our astonishment
in ascending the day before, proved to be merely
some of the bigger lumps of these dark lava torrents,
carried by their mass, a little further than the rest.

Cooling and hardening on the surface, cracking,
breaking up, and falling forward in clinkery masses of
rattling cinder and stone,—such, must have been the
mode of progression of these black streams, as with
many of the Vesuvian lavas under ocular inspection.
The material of them,—always the rough granular lava,
with crystals of white glassy felspar mixed up in it,
like chopped straw,—has a strong tendency to form
in spherical shells, or in a veined curving laminated

mass; the denser layers being partially separated by
somewhat imperfect and frothy portions, like an
incipient pumice.   What with the varying effects of
refrigeration, of pressure on the once pasty, viscous
masses, their rolling, falling, and tendency to crystallization and cleavage, in the good old days, when these
cascades came roaring and burning down,—the forms
of the blocks are the most varied and grotesque that
can well be imagined.   (*See Photo-stereograph, No.* 9.)

Close round about our walls, there were rocks so
curiously fashioned into wild dream-like pictures of
contending beasts, that I have had an earnest request
from a sculptor, for a copy of one of the photographs
of them, to work up into a group.   The bear, the fox,
the baboon-faced fish of Simon's Bay, a short-nosed
alligator with the projecting eyes of a prawn, can be
pointed out now by a child in these faithful portraits;
and were yet more striking, as we viewed the rocks
themselves, when living amongst them morning, noon,
and night.

Along the tops of the ridges, against a clear sky,
the almost sculptured figures were thrown out in
powerful relief.   There too, the idea of motion of the
whole stream was brought out so remarkably, that
one could almost believe in the existence still of an

under current of molten lava, keeping all the hot slags above it, alive and in movement. We could almost fancy we heard them grinding and crushing as they rolled heavily onwards; carrying their tortured freight of Roman emperors and Don Quixotes, old women and stout aldermen, writhing and twisting, and groaning downward, helplessly on their Tartarean bed. No hope seemed there for any of them; no rest for a moment; nothing but a constant, cease-less, irresistible progression in a community of pain. The fat man, and the lean man were bowed down together, in their unutterable fate; and a late well-known Lord Chancellor, with his nose up in the wind, was there, flat on his back, and going head first, down the inevitable cataract of blackened crags.

In some of the sandstone formations of South Africa, wearing away under the influence of weather, into a thousand fantastic shapes; where there ap-peared the similitude of birds, and beasts, and plants, and ships in full sail—I have yet seen nothing like the general uniformity of purpose that seemed to prevail amongst the semblances, that would perversely force themselves before us, from out of the chaos of these black lava torrents at Alta Vista.

Let me add, however, that there was one redeeming

figure; that of a Spanish nun, on her knees, with her
hands joined as if in prayer, and robed in the pendant
mantilla. On exhibiting to friends, a photograph of
this part of the lava stream, magnified and depicted
on a screen by aid of a Drummond light, they
could hardly understand that " the nun " was not,
either a real statue ; or my wife personating one.

Utterly unscientific as such features may be, we
could not help noticing them, and wondering at them
too ; but more were we surprised still, to learn, that
up amongst those loose stones, was the water place
from which our supply was to be drawn ; the celebrated " ice-cavern" of many writers. Broken stones
can nowhere form a good cistern for holding water ;
and least of all, on the side of such a hot, dry mountain as this cone of the Peak. Excepting the little
water-spot at Guajara, and that was a long day's
journey off, we had not seen a drop of the precious
fluid during the whole of our ascent ; a more arid
tract could hardly be conceived, and arid in a great
measure, on account of the porousness of its pumice
soil. Now, however, that we had arrived in a region
where the porousness had increased to the extent of
loose rounded stones on a steep slope, — we were

told that there, above our heads, was a fine tank of water.

Much as I had wished to go up and look into this wonder for myself, I was kept at the station by the term observations. Moreover, our excellent assistants, the sailors, had got into trouble; they had been living so freely on goat's-milk cheese and the richer goat's-milk, taken exclusively, and *ad infinitum*, that they now required to be medically treated. They came out of it very subdued, and declared most innocently, they would never touch those condiments again.

Meanwhile the Spaniards ascended to the ice-cavern, and brought down splendidly pure water, and a large lump of hardened snow. The latter served a most epicurean purpose. A portion of the milk in a glass flagon, that in the long journey on horseback was reported to have become undrinkable, because it was not only sour but thick—proved on examination to have been involuntarily churned into excellent butter. As white as snow was this specimen, but that is natural with the product of Teneriffe goat's-milk; and butter of any sort being a rarity in the land, my wife treasured up our unexpected supply, and preserved it sound and hard, notwithstanding the heat of the sun by day, through means of the portable cold brought down from the ice-cavern.

High as we were now above the zone of *retama*, and above every plant except an occasional lichen on a rock, there was little fear of any of our men again getting surfeited with goat's-milk; for we were quite out of the range of Arcadian visitation. With the absence too of plants, there was happily a similar scarcity of insects; we saw in fact nothing moving, except occasional long-legged spiders (*Phalangium spiniferum*) at night.

Harmless spectral things they were, and their legs very long; so long, thin, and filamentous, as probably, in the creature's absence of web-spinning power, to enable it to catch a fly by entanglement with its legs alone, if there had been a fly to catch. But we did not see such a thing anywhere for them; and they wandered about at night like troubled spirits over the roof of our tent; or occasionally stopped short, turned round, and seemed to take a mournful and astonished look at us; but pirouetted once again and toddled off quickly enough, the instant a motion was made towards them.

The immediate soil of our encampment was a coarse

sand of broken pumice, whose parent rock cropped out a few yards in front of our walls, as a sheet of rough red-brown stone, with a long vesicular sort of grain, looking like petrified wood. A few yards further still, there was a break-down of the crust, giving a famous sectional view of the composition of this class of lava exudation; and here was obsidian *in situ* at last. Not, however, the pure obsidian for manufacturing Guanche edge-tools; for the present, though as glassy and brilliant as it could well be, had small interspersed white crystals of felspar, that sadly tended to split the mass to pieces, and render it friable.

The whole thickness of the bed was but three or four feet; and of this, the lower third part only was of dense material; for vesicles had formed in the upper portion, gradually enlarging as they reached the surface, until it had the appearance of a well-raised loaf, somewhat pulled out however in the direction of its downward flow. The red-brown colour at the top, seemed to be merely the effect of oxidation, the true colour being black; though as the partitions of the hollows became thinner, their translucency caused them gradually to change from black, to the green of bottle glass; and every now and

then, one of the white crystals of felspar was found
entangled and unchanged.

The material indeed of this obsidian stream, appeared to be the same as that of the later and super-
imposed ones of stony lava; but in place of having
exuded, as they had done, barely melted,—the older
ejection had rushed forth perfectly fluid, spread
far and wide in a thin sheet, and being rapidly
cooled, had assumed a glassy, in place of a stony,
texture.

Original heat was further testified to, by the
pumice of some previous eruption; for this, the obsidian had overflowed, positively burning the sponge-
coloured masses with which it had come in contact,
though they were not carbonaceous. On raking out
some of these blackened lumps, with white ash hanging about them, from under the obsidian bed,—we
could hardly but believe that they were cinders of
some traveller's fire of the last week, so natural and
fresh did they look.

Everywhere round about us, traces of fire were
predominant; our walls, built of half lava, half
pumice, were quite a volcanic museum in themselves.
Hence perhaps my thoughts had probably a pre-
judiced bent, when, going out at night to view the

stars, a vivid white light on the floor of the crater caught my attention. It had no particular shape or sensible size, it was simply a centre of light; but so intense, that it appeared to the eye with great rays like a giant star, or like the sun as it must be, when viewed from the orbit of Neptune.

Each and all of our party were astonished; and the place of the appearance was close to one of the " Druid-circle" craters. Had the ground cloven there afresh, revealing the incandescent flood below?

A little "finder" telescope would explain nothing; but the Sheepshanks five foot being brought out, showed—that there was merely a burning of some *retama* bushes going on.

The light of such a conflagration is generally remarkable for its redness, as well as rambling extent; but the present display exhibited a striking whiteness and intensity. These qualities seemed to be owing, to the bushes standing on a steep slope, where the fire could leap up from one to the other; and to their being the arid growths of the crater soil; while the flames, condensed by distance, were undimmed in their passage through the thin transparent air, of this land above the clouds.

CHAPTER III.

BRINGING UP THE TELESCOPE.

NO very long residence at Alta Vista was needed to assure us, that this was the place, the centre as it were of the lava streams, defended by them from the winds, North, West, and South; but with a clear zenith view, and enjoying the most pellucid of atmospheres, or rather so little of any sort of atmosphere,—without question, this was the place, whereon to erect our friend, Mr. Pattinson's large Equatorial. True, that experienced local men had declared the bringing up of such a telescope to be impossible; but as the interests of science demanded that it should be brought up, I determined to descend the mountain, and see how the affair was to be managed.

There was plenty of useful work meanwhile for our men to accomplish at the station; the barewalled spaces having to be roofed in, and converted into habitable rooms. Towards this end, a large quantity of patent felt, and bundles of young fir poles had been procured in Teneriffe; in addition to certain beams, plates of glass, shutters, and door-hinges which had accompanied us from Edinburgh. With these materials, the sailors and Mr. Carpenter set zealously to work, on the morning of the 25th of August, just as my wife and I were leaving the station, with three Spaniards and three horses; of whom, one carried a portable meteorological observatory. It was rather a curiosity in itself, this observing case, having been constructed during the last two or three days by our yacht carpenter; it consisted in a tall box, with ventilating open-work sides, and all its external surface covered with new tinfoil, shining like silver, to preserve the contained instruments from solar radiation.

Away we went then in the cool of the morning, and more easily than had been expected, as our two active Spanish servants, Manuel *dit* the Marquis, and Manuel the Guide, had been voluntarily improving the road, and making several long inflections, where experience had proved, during the ascent, that a more gentle slope was necessary. From our barren and desolate neighbourhood of Alta Vista at 10,700 feet of elevation, we now glided quietly down between

the black lava ridges, right and left of us, to the
first *retama* bush at 10,000 feet; and then to
"Estancia de los Ingleses," 9710, where all the
meteorological instruments in our observatory with
the metallic lustre, were read and recorded. This
process was afterwards repeated nearly every quarter
of an hour through the day; rather annoyingly
perhaps to the animal who carried the brilliant burden, for he was slewed about unsparingly on the
steep sloping surface of the hill, until the barometers
were vertical, and the door of their enclosure in such
a position, that it could be safely opened to allow of
the reading being effected, without letting in too
much light and heat.

The sky was bright, and the sun already scorch-
ingly hot; but the S.W. wind blew with refreshing
coolness, and at a steady rate of about six miles an
hour. As the horses paced the regular path downwards to Montaña Blanco,—I descended on foot
the long strip of loose pumice, cracking with my
hammer every bit of obsidian I came across, and they
were many, in vain search for the pure and dense
material of Guajara. The fractures were eminently
glassy, often splendidly iridescent, but always con-
taminated with the white felspar crystals of Alta

Vista obsidian. The ends of many streams, such as we saw there before our door, cropped out on this slope, and had been more or less broken up by nature. Obsidian, vesicular lava, pumice and all, they now rolled downwards, comminuted and mingled, but still offering famous mineralogical specimens of the penultimate exudations of Rambleta. The ultimate, being always, blue-black, granular varieties of lava.

The nakedness of these last, was as remarkable as the abundant manner in which the ante-penultimate had covered themselves with pumice. Montana Blanco was a mighty tumulus of this sort; and deep under its surface of fragile tender little specimens of yellow stone in filagree, that looked altogether at a distance like a harvest ready for the sickle,—there must evidently be concealed some hideous rents and chasms of volcanic fire. From these have been emitted the red lava streams, that have pushed their way through the whole superincumbent mass of pumice; and are now seen flowing down its southern slopes, in a viscous, wrinkled manner, as already described.

Working along first eastward, then northward, we threaded the red rocks of the crater floor, easily enough in this quarter, thanks to abundance of

pumice; and at last arrived at the Canada del Icod,
or one of those narrow sandy flats between the out-
poured subaerial lavas on one hand, and the walls
of the great crater, submarine at its period of formation, on the other. This portion of the circular
rampart, is situated west of the gap by which travellers usually ascend, from Orotava; and being about
180° from Mount Guajara, enabled us to compare
opposite sides of the ring.

The northern wall then, it must be confessed on
making such comparison, is far inferior in height to
the southern, but is precipitous, and even tends to a
basaltic formation, along its summit; while in its
lower portion it is interesting to find a stratum of
hard, but brittle greenstone, breaking into half-
crystalline fragments; exactly as with the similar
stratum near Guajara. Still another circumstance
well worthy of note, is in this north portion of the
great crater wall, being analogous to that part of
Somma, which is supposed to have been destroyed
and scattered in dust and ashes, to form the winding-
sheet of Pompeii, on that disastrous night when the
cone of Vesuvius arose. Teneriffe has also tried the
strength of its seaward or northern wall, and destroyed a large portion of it; but the paroxysm we

may hope is now quite over, its Vesuvian cone complete, Orotava safe, and this invaluable fragment left standing, for a permanent record of the majesty, and structure of the volcano, in ancient troublous periods of early natural history.

Ascending the Icod cliff, at a low fractured bend of the rampart, and passing strata of clay, with a bright scarlet surface, where in contact with lava beds,—we at length clambered up to the summit; and then began to descend leisurely a long external slope, smooth, broad, and with a dip of only 12°.

Here we exchanged the S.W. wind, that had been blowing on us all day, for the N.E.; but the sky remained blue, and the solar radiation excessive; the aridity of climate, too, was unsoftened; and still the same one and only plant, *retama* or *cytisus nubigenus*, continued to decorate the soil. From Estancia de los Ingleses, where we entered the zone of vegetation, down cindery slopes to Montaña Blanco; and from Montaña Blanco through miles of long winding passages in the crater floor, nothing but *retama* had greeted our eyes. Now too that we had passed the crest of the wall, and at an elevation of only 6700 feet were descending its external flank, there was still nothing to be seen but the solitary bush here, and bush there, of this remarkable broom.

Had we dismounted from our horses and searched in rigid botanical style, I doubt not that we should have found several specimens of stray little plants; but so few, as not to constitute a millionth part of the weight of *retamas* passed in the same space of time; and forming therefore, no sensible portion of the real practical vegetation.

At their highest limit, the plants were flat, and trailed their hard wood stems along the ground; but they had gradually improved as we descended; until now, we found them ten feet high, fifteen broad, and with stems more than six inches in diameter. The soil, a rather friable mixture of sand and clay, as if from the decomposition of green-stone, was deep, and agriculturally, I should have said, good; so good, that it was strange indeed to see, nor corn, nor grass, nor weeds, nor anything in fact, but here a *retama* and there a *retama*, whether one looked up-wards or downwards, or on either side. In the spring time, when all the bushes are white with their sweet-scented flowers, this region must be the Arcadia of goatherds; though now, the constant sight of the blue grey twigs grew so wearisome, that we were

quite relieved when the shouting of two of our
muleteers in front, announced that they had suddenly
caught sight of the gardens and vineyards of
Orotava.

These men were standing on the edge of a precipice, whose line ran nearly north and south, or in the
direction of a radius of the great crater; and it was in
truth, the western side of one of its principal ravines,
or fractures, formed at the time of forcible elevation.
Except near the summit, the eastern or opposite side
of the ravine was hardly distinguishable, so immensely
was it removed into distance, by reason of a remarkable subsidence of the ground in all that quarter.
This indistinct sort of Ætnean "Val del Bove," gave
to our view in front, a fine semicircular series of
highlands and serrated cliffs, seamed with dykes that
glittered in the sun;—and in the foreground, or
below,—the depressed amphitheatre of the valley of
Taoro, the fertile seat of Puerto and Villa, Orotava,
as well as many towns and villages besides.

Looking down from our arid heights some two
thousand feet in depth, to where torrents on torrents
of lava, the early outpourings of the cone of eruption,
have chased each other in headlong flow from the

gate of that great crater above; and where they have
been followed since by winter rains sweeping down in
annual cataracts, and cutting out ravines that were
fearful even at this distance,—we stood on the parallel
as it were of a lower region, and a more fertile climate.
Mere perpendicular height, without any alteration of
distance from the mountain top, now alone separated
us from rich green vegetation, fields and orchards and
the abodes of men. All these we saw beneath us, as in
a panorama, whenever portions of the land cloud or
" vapour foot" of the mountain, opened or closed.
Closing, they appeared to cherish the green slopes with
affection; and opening, to show with pride the rich
results of their love.

We could not, for steepness, however much we
wished, descend the almost vertical cliff of this rem-
nant of the crater flank that we were upon, by name
Mount Tigayga,—for us, there was nothing but to
plod straight on, as at first, along the frontal face, or
slow descent towards the sea, and to *do* our *retamas*.
So we settled again to our task, soon lost sight of the
fertile land below, and were contented per force once
more, with merely the variety of a big *retama*, and a
little *retama*, and a little *retama* and a big one over
again.

Half an hour's more patient tribulation, producing incipient slumber; and then, in a moment, we were positively startled by the appearance of a new plant! Had it been a tiger or an antelope, and we on sporting bound, the effect could not have been more like magic.

A very wee thing was our new plant, growing almost flat, like a creeper amongst the stones, but of such a vivid green. From Alta Vista down to this point, had been nothing whatever seen to compare with it. Thirty seconds more plodding downwards, and another of these bright-green plants was seen; then another, and it was a little taller; then another and another, and in three minutes no other plant than this was before us.

We turned round immediately; and, a gentle rising in the ground, concealing the immediate distance behind,—there was not a single *retama* in sight on the whole of Mount Tigayga! Only on a distant slope of the Peak westward, could be seen where the grey of the broom came down to, and was sharply separated from, the green of the heath, which our new friend turned out to be; while on one far-off ridge, a solitary old pine tree was growing, exactly on the line of demarcation.

Less than five minutes of slow walking had thus

sufficed to take us out of a region of nothing but one peculiar plant, and convey us into another, peopled as exclusively by a perfectly different tribe. Here was botanical geography made easy indeed! and if all the world were like this upper part of Mount Tigayga, what accurate trigonometrical surveyors would not botanists become.

As we penetrated further through the region of heaths, they grew more upright, and in fact soon claimed to be considered quite sprightly little shrubs; yet they ill occupied, or filled, the ground. Each individual was far separated from every other, with only the bare red clay between; what with this, and their green foliage so vividly bright, one could hardly fancy but that the place was some gardener's nursery, where each precious plant had been set out carefully by itself. No signs however of human agencies yet.

In a quarter of an hour more, a fern was seen; then another, then more, and the zone of ferns was fairly entered. They were found to be perfectly sociable with the heaths; for they grew in juxtaposition with them, and both increased, at every step we took, in size and beauty. Some waving fronds of the former were four and five feet long; and the heaths, all of one species however, formed exquisitely graceful bushes with spiry, upright branches more than ten feet high.

In half an hour more, a laurel tree was passed; then a bay, then a group of them. The soil was now a friable loam, worn into deep goat paths, whence the horses' feet sent up clouds of suffocating dust; but looking down, some few grasses were seen under the now complicated upper growth of heath, and fern, and laurel. Presently with more and greener grass, the yellow flowers of the hypericum enlivened the collection; and at last about 2 o'clock P.M. we turned down to off-saddle in a shallow ravine, where was a spring of water, trickling out from under the gnarled roots of a patriarchal laurel tree; and where the grassy banks were soft and turfy, with space and shelter, wild flowers and green leaves, for the enjoyment of both horses and men.

The elevation of this delicious resting-place, was 3400 feet above the sea, and seemed to be but little higher than the upper limit of the land-cloud, portions of which were floating near. The air felt steamy, and luxuriously oppressive. Invited to enjoy ourselves by the gentle genius of the place, we reclined at our length on shaded sward, and by the side of the bubbling fountain, discussed a grateful luncheon of cold tea and biscuits, choice bananas and luscious grapes; some of the few, that the fell vine-disease had spared, in the gardens of our kind Orotava friends.

To perfect the amenities of the scene, the human element was not wanting; for troops of children surrounded some women, who were picturesquely employed in washing clothes by the stream, a little way below where we had located ourselves. These children presently made up to us, and brought my wife some charming white mosses, whose filaments were like frosted silver.

Everything in short, in this delightful retreat, seemed to combine in one end; and that was, to produce in our minds, for the time, perfect oblivion of black lavas and red; obsidian, pumice, and trachyte; and of all things which appertain to the upper regions of the Peak.

Soon after three o'clock, we were again mounted. The surface of the land-cloud appeared now close by, but was thin and broken; grasses abounded, together with heaths; a little further on, we came to fields of corn and lupins; to a farm-house, and blackberry brambles. A puff of mist enveloped us for a moment, and immediately after, at 2400 feet of elevation, a fig-tree presented itself, and was quickly succeeded by an American aloe, and by Indian corn.

So far along the *sea-face* of the slope of Tigayga; but now we turned into a winding and paved mule-path, that led down its steep *eastern side;* passed more houses; passed countrymen with mules kicking out in most dangerous places; passed under magnificent escarpments of rock; and presently looking upwards,—saw the cloud now above us! Passed vines, pumpkins, and cactus; passed two streamlets of water, and finally,—through long garden walls decorated with small rich ferns, and overshadowed by pear, mulberry, and peach trees,—we entered the town of Realejo, only 900 feet above the sea.

Strange sight and melancholy, to see a large town approachable only by mule-paths. The Spanish government, however, has lately commenced a fine broad carriage road from Santa Cruz to Orotava, eventually to reach this place, and ventilate the closeness of its ancient atmosphere, social and physical.

Better adapted, at present, to bring out the beauties of plants, than the talents of men; the neighbourhood of Realejo pleased us most, with its rich

groves of oranges, lemons, and bananas; fig-trees
heavily laden with their purple fruit, and an occasional gaunt dragonier. Evening came on soon after;
and in the darkness that followed, we only noticed
one change more in the vegetation,—viz., the appearance of euphorbiaceous plants, before we touched on
the coast.

No great time had elapsed next day, ere we held
a council with our invaluable friend and abettor,
Mr. Goodall. The three huge boxes into which Mr.
Pattinson's Equatorial had been divided and packed,
by its maker,—were undoubtedly too heavy, to be
taken up the mountain by mules. Had it been otherwise, great convenience would have followed from the
arrangement; for the parts to be put together in
order to reconstruct the instrument, would have been
few and simple. Go one step further, the optician
had told me, and you will be bothered exceedingly;
for the several pieces in which one of the first-class
equatorials is constructed, are manifold in the extreme. To divide the instrument still further, was
however the only solution of our difficulty; and with
the assistance of a very clever German watchmaker,
settled some years in the town, we commenced operations.

The German produced his screw-drivers, and the sight of them established him in my good opinion at once; there was a cut about them, an excellence of original make, combined with recent improvements of his own, which showed that Herr Kreitz was a genius, and that he was as fond of, and as skilful in, mechanics generally, as in watchmaking in particular. Moreover, he had brought a young carpenter educated by himself, and a trusty old hand besides; while Mr. Goodall, in whose merchant's yard we now were, contributed two or three men, with several mattresses, tables, and boxes.

As we unscrewed the cases, Dons of Orotava, present by special permission, also clustered around. Every one of the party, even down to the lowest porter, took a happy interest in the proceedings. Ejaculations were made as the contents of one box after another were disclosed; and when at last the two half tubes of the telescope in polished brass, with finder, micrometer, &c. &c., were discovered, there was a general gasp of "*instruménto*," with a sort of superlative accent of admiration and astonishment.

Attacking first, the heavy cast-iron portions of the stand; these were separated, into their ultimate atoms as it were; axes pulled out of sockets, circles

T

removed, clamps taken off, microscopes unshipped, until the convenient mattresses were covered with innumerable " *disjecta membra*." Then came the planning of new boxes; each heavy fragment in a box by itself; the smaller pieces, many in one box, but each to have its own particular partition, to keep it from falling against the rest. Eventually, the contents of the three original chests, were distributed into thirteen new ones, and the good old town was ransacked for screws to fasten on the lids. Plain nails were to be had in abundance, but good screw-nails seem to be bound up with the march of Anglo-Saxon civilization.

On the 30th of August we started again for Alta Vista, with the Pattinson Equatorial now safely packed up in its new boxes, and mounted on seven strong horses. Straight we went up the valley of Taoro, stopping only each quarter of an hour for meteorological observations.

The land-cloud was met as before at 2500 feet of elevation; and above it were the fierce sun and arid air. Up along the sides of rugged ravines that we had overlooked from Tigayga, did we now proceed; and over old sheets of lava,—where the stony soil enabled the *codeso* to appear, and interpose its zone between the heaths and *retamas*,—until we entered the basin of the great crater, at its north-eastern gate, about noon.

The heat, or rather the radiation, was terrific at one point of the floor, where the air was perfectly calm. On passing this, we found a S.W. wind blowing on the further, and upper side. It soon blew strong, and not refreshingly; the horses were continually flagging, and on reaching Montaña Blanco, we were obliged to make a long halt to rest the exhausted creatures, before we could think of climbing the cone of eruption.

Meanwhile the wind increased in strength, the sun declined, the air rapidly cooled, and a cloud formed on the top of the Peak. Rather a portentous cloud was this; not very large or heavy, but so furiously impetuous in its motions. Of cirrous structure, the filaments of mist formed, dissolved, and formed again, and went driving almost with the speed of lightning over the mountain-top, twisting and writhing as they fled along, like snakes of Medusa's head. The thermometer was presently found to be falling momentarily, and the dryness of the air was decreasing in an unprecedented manner, so not a mo-

ment must be lost; our horses accordingly were
reloaded, and the ascent recommenced.

Long and very weary, with innumerable stoppings,
was our way; then, beating the horses, and shout-
ings of " *andār cabǎllo*," with addition of bad names
to many of them, followed; and with no great success, though some of the names were very bad indeed,
even amongst Spanish maledictions; yet by steady
perseverance, Estancia de los Ingleses was reached at
last.

A short rest there, and then a steeper slope was
entered on. The horses were too tired to show their
favourite Canarian vice of kicking; and the weakest
of them were got over the most difficult turnings by,
one man pulling a-head, another pushing astern, and
others clapping on to any side that they could get at.
In this way the worst of the ascent had been overcome, the station-walls were in sight, and our men
coming down to help,—when one of the horses fell!

He fell sideways down the slope, made a complete
somersault, and then tumbled against the next horse,
who was carrying our meteorological observatory.
He too went right over on his back down the steep,
turned up on his legs again, and after several struggling bounds succeeded in stopping himself. Mean-

while the first horse was lying on his side amongst sliding-stones, and only kept from going further, by men holding on to his bridle; two boxes were still lashed fast to him; while a third, carrying, unfortunately, the clock-work motion of the equatorial, had got detached; and in the midst of a shower of cindery blocks, went bounding down the mountain-side.

CHAPTER IV.

BATTLE OF THE CLOUDS.

THE sailors expressed themselves delighted to see us again, and we were as pleased to witness the results of their active handiwork. There were now five rooms roofed in, round about the telescope enclosure; while along its inner southern side, ran a verandah, intended for meteorological instruments; and since discovered by the carpenter, to be a luxuriously calm place to sit in, when there was wind everywhere else.

This was previous to our return; for subsequently, the western gale grew so violent and squally, that there was no rest or defence for any of us. We were told, that a change had come over the weather during our absence, cold and dark, as if autumn had arrived. We looked to the thermometer and found it, on the morning after our return, 18° lower than when we had set off for the telescope; the air too was undoubtedly thick and turbid with dust-haze; but the chief change that had occurred, was in the appearance of the sea of cloud. Not only did this vast surface seem to have received a terrible shake, that had broken up its accustomed forms of long rolls of cumuloni, into short cumuli, but—an enemy had appeared in the field.

If there had been generally any part of the sea not amenable to influences of north-eastern wind, and not covered in by those clouds, it was to leeward of the Peak; there, long extents of water surface had been occasionally seen from Guajara. These were now precisely the regions that had become filled with a strange *south-west* cloud. Cumulostratus in character, and at the same height above the sea as the north-east stratum, this new cloud advanced to the charge, its masses of mist hurtling on over each other, as though they were eager for conflict.

The first shock seemed to take place near Grand Canary, whose tortuous ravines and bristling peaks offered many vantage grounds for fight. The advance guard of S.W. cloud rode high up that side of the island, well supported by its main body on either flank; but the N.E. cloud was not to be disturbed, that time, from its strong position. Then filling up all the space behind Guajara, the invader charged successfully up along the southern coast of Teneriffe,

carrying all before him as far as Santa Cruz. Next
he tried to cross the back-bone of the island, near
Laguna. Laguna is sadly fated to be a battle ground
for clouds; as, modified by the form of the island,
three aerial currents ordinarily meet there, the N., E.,
and S.W.; and the result of such conflict, is generally,
a fall of rain. On this occasion, all petty local
animosities were merged into the grand contest of
N.E. and S.W., which now with equal forces con-
tended by night and by day, for the possession of this
cloud loved spot.

It was, in sober truth, most exciting to watch the
varying success of the combat, that was going on,
really—high in air; but for us—far below our feet; and
to witness the science with which it was conducted. A
gallant band of the S.W. sallied bravely up the hills,
seized the summit and were having it all their own
way, when a rush was made by heavy columns of the
N.E. They recovered the lost ground; but being
nearly expended in so doing, and their weakness discovered by the enemy, he came on quickly from his
main body, in overpowering numbers crowned the
ridge, and began pouring down the opposite side.

This was too dangerous a success for the N.E. to
allow, so it moved up all its masses, and by a well-

directed attack at some weaker part of the enemy's line, compelled him to call in his troops towards the centre, in vital defence. At other times the N.E. tried to push out his opponent by main force; and then, if well balanced, they both, from their slightly inclined opposing forces, rose high vertically in the air; a conspicuous example to the multitudes in either army; and dissolved away in the contest, or fell over, maimed and mangled, to their respective sides.

All day long this engagement had been waged, and on the whole with such equal success, that we could not say by the evening which party was in the ascendant; but our hearts were with the N.E., for a predominance of that current, was the symbol of summer.

On the mountain at this time, the west wind was blowing so violently, that the pumice-stone sand about us, drove along in clouds; and so cold were the gusts, and so mischievously inclined, that we had to employ a great part of the day, in improving our dwellings: putting large stones on the roof, to assist its few nails in keeping down the felt; and filling crevices in the dry walls, with bits of wood, cloth, paper, wedges of stone, and whatever we could get to stick in.

September the 1st, a term-day, found westerly wind as strong as ever on the mountain; the combat between the lower cloud strata, undecided; and a large icicle on our water barrel. The clearness of the air had however improved, and on taking a black bulb thermometer to the shelter of a breakdown in an obsidian lava stream, the mercury rose so rapidly as to be above 180° at half-past nine o'clock A.M., the temperature in the shade being only 49°.

There was so much sand flying about in these squalls, that we could not venture to open the boxes of the telescope; but we built up and erected the pier intended for it. A foundation of beams was first laid in the ground with lime; and an upper structure added, of a box figure, formed of planks arranged by Mr. Cooke, the rising optician of York. It was about nine feet high, five feet broad, and one foot thick; and being filled with stones, acquired the inestimable firmness of weight. In the evening, we were happy to observe that the S.W. clouds, although still very extensive about Teneriffe, had been fairly driven off Grand Canary by the N.E.; and we earnestly hoped, that victory might soon declare itself similarly in our own neighbourhood as well.

Photo-Stereograph. II.

CLOSE VIEW OF ALTA VISTA OBSERVING STATION FROM THE EAST.
ALTITUDE 10,702 FEET
p. 326.

Printed by A. J. McKnight under the superintendence of James Glaisher, Esq. F.R.S.
and published by Lovell Reeve.

After a sharp night, when the temperature of the air went down to 40° and that of evaporation to 28°, the sun appeared bright and clear. A black bulb thermometer being tried about nine o'clock A.M., rose in a very short time from 48°, to much above the top of its scale, or to more than 180°. What the full temperature must have been, had our instrument been capable of marking it, were perhaps rash to speculate on; but the quantity certainly recorded, being quite sufficient to show admirable qualities in the atmosphere for astronomical observation, and the wind having much decreased,—our unpacking and reconstruction of the equatorial began.

We had already extracted certain parts of the polar axis, its frames and bearings, and were putting them together on the ground, all of us on our knees in a circle round about,—when crack went something, with a flash before our eyes, and a blow on the ground as if a meteorite had struck it. The poor interpreter, who had been usefully holding one of the circles that was being screwed, jumped up with a hand swelled to double its proper size. In a moment of perfect calm, one of the largest stones on the adjacent roof, had slipped off without any warning, and falling into the midst of us, had struck his unfortunate hand, producing an injury more painful than the fracture of a bone. He bore it admirably, and not long after, with his arm in a sling, was again in the telescope-square.

Our work once more progressed, and prosperously. All present, even including the two Manuels, seemed to take a personal interest in the instrument being erected; and what with the ready mechanical intuition of the carpenter, and the tower of strength which the second mate proved to be, in raising sundry heavy masses of metal, up to positions they were destined to occupy—the work was accomplished before sunset.

With night, came on clouds; cirri and cirrostrati high in the air; and renewed conflicts of the strata below. The S.W., we grieved to find, was again audaciously invading the territories of N.E.; they met in conflict over Laguna, and elsewhere invaded their respective sides of the great crater. There was even a threatening of rain, and we had to protect our newly-erected telescope with waterproof canvas.

Next morning affairs were not improved, so that we could do little else, than chronicle, with photographic camera, the state to which we had arrived. (*See Photo-stereograph, No.* 11.) Right in the foreground, on the brilliant pumice-stone soil, may be seen one of the tumbled masses of lava from above : a little further off, on the right, is a portion of a far larger similar mass; while on the left, stretches away an angle of our architectural labours, showing a cosy corner between two rooms, whose flat roofs are concealed by perspective.   Over the top of their walls, towards one side, may be distinguished, the telescope; its eyepiece end turned up, and covered with thick white cloth; the declination circle and axis, with painted canvas.   Beyond and above all this, is seen the upward slope that leads to the Peak; the brighter part, a red lava stream; the darker, one of those strange torrents of blue-black material, exhibiting only a long embankment of broken cindery stones.

Climbing up one of these ridges, that on our north, a famous bird's-eye view of our whole station was obtained, showing all its flat roofs, the open interior of " telescope square "; and, above all, the tremendous black lava stream towards the south. (*See Photo-stereograph, No.* 12.)   In the square, the carpenter seated on a box, is busily hammering; while in the distance, lying on the ground within a rude fire-place, built at the foot of the lava bank, is the patient interpreter, wrapped up in a blanket cloak, and suffering, in silence, from his unlucky hand.

In the course of the night, notwithstanding many clouds, we were able to get observations for necessary adjustments; and, more important still, the clock-work motion was successfully applied. Its roll down the hill, had only broken a plate-glass casing, and was replaced by our carpenter, with panels of deal. This man's Robinson Crusoe genius furnished me, likewise, with an observing chair for zenith work, formed out of some of the packing-cases and bits of iron hoop.

To another carpenter, I had also been greatly indebted, in Edinburgh, for his continued attention in making, filling, and packing all the boxes for this Teneriffe expedition; and for his provident care in thinking of, and supplying, many little odds and ends required on such a service. One case that contained a nest of boxes, each with a plate-glass cover, intended to try Sir John Herschel's method of economically utilising solar heat,—seemed to this good man to be the crowning point of his labours; for when we opened it on the mountain, there was a piece of paper with the following inscription:

" Rankin Lawson, joiner, Edinburgh, packit this box, I hop it will arrive safe."

On the morning of September 4th, a great change

presented itself in the lower clouds; an improvement we of course thought it, because the right party for us, had got the upper hand. The N.E. had indeed gained a great victory; and its innumerable regiments of white rollers, were now filling all the sea east, as well as north, of Teneriffe ; while the forces of S.W., were everywhere retiring before it. By the close of the day, only a few small dispirited cumuli of this iniquitous party were visible; and they were seeking protection ignominiously, under the southern side of Teneriffe.

That night there was no wind; not a single upper cloud, and the definition of the air was admirable. With all due precision, aided by the refinements of its equatorial mounting, the large telescope was now turned on one test object after another, amongst double stars; and magnifying powers employed, from 160 to 800. To the credit of instrument and atmosphere, these high lenses were borne perfectly. Stars of the 16th magnitude as " a " of a^2 Capricorni, and b of β Equulei, were seen without difficulty; and pairs, only a fraction of a second apart, as ε Arietis, λ Cygni, and γ Andromedæ, were separated. In fact all the highest tests that were then known to me, either as to brightness, for a proof of the trans-

parency of the air, or as to closeness, for a proof
of its steadiness,—were transcended on this admirable night. Two more nights of the same high
quality followed, and enabled many useful observations to be tabulated, with the same general result as
above.

To make the experiment complete, there ought to
have been similar observations procured at the sea
level, with the same eye and tube. Such I had
attempted, during three different visits to Mr. Pattinson, of Newcastle, the hospitable owner of the telescope,—but was defeated on each occasion by clouds;
low clouds too, as I believe, not above 3000 feet in
altitude. And by similar clouds would any one in
Orotava or Santa Cruz, have had all view of the
heavens shut out on those three evenings, which
proved so splendid at Alta Vista, only because it was
raised more than 10,000 feet above the plains.

Whatever therefore the experiment wanted in nice
completeness of comparison,—between stellar light as
seen on the mountain, and again at the sea level,
after undergoing the absorbing and deteriorating
influence of an additional depth of 10,700 feet of the
lower atmosphere,—was made up by the fact of our
having been only enabled on these particular nights

to see anything at all of the stars, through dint of elevation of site.

Over and above, however, merely seeing the stars, I could not but perceive that the degree of their definition was far finer, both over a large extent of sky, and for a greater portion of night, than had ever fallen within my experience at lower levels. The worst period of the twenty-four hours for astronomical observations at such places, is invariably during morning twilight; and the worst part of the sky at that instant, the eastern region. Even in the fine climate of South Africa, both stars and planets so circumstanced, have been seen to lose their definition in the telescope to such a degree, as at last to become little more than amorphous balls of luminous hair.

Yet it was precisely at such a time, and in such a quarter that I was observing Saturn, upon Alta Vista, with a magnifying power of 500 ; and the general impression left on the eye, was,—the remarkable sharpness of the edges of planet and ring. Owing to this circumstance, the fine division of the outer ring, a much disputed point, came out with singular distinctness. Not much, however, could be ascertained, beyond what was already known, with the planet comparatively so unfavourably situated. With Jupiter near the zenith, the case was very different.

The usual mere streaky bands which cross his disc, became resolved in the telescope, under high powers, into regions of cloud. The brighter spaces were the clouds, and their forms were as characteristically marked, and were drifting along as evidently under the influence of a rotation wind, as the cumuli and cumulostrati which the terrestrial N.E. current, was, at that moment bringing past Teneriffe, before our eyes, and under our feet. On each of three nights that I made drawings at the telescope of these Jovian clouds,—the effect of the planet's rotation was abundantly evident; while in addition to this, there were minute changes in the relative positions and forms of the vaporous masses in either hemisphere, that indicated as well the presence of winds, as the ephemeral nature of mist.

Far more striking, however, was the testimony borne by the more constant forms of the cloud, seen best toward the equatorial part of the planet. At this tract one could not gaze long, without acquiring the impression of looking at a windy sky; the whole zone of vapour seemed to be in motion, while from its ragged edge, portions were torn off and were driving along, some of them rolling over and over, and others pulled out in length and rearing up towards the fore-part, like a sailing-boat scudding before a gale.

Owing perhaps to the effects of perspective, the polar zones appeared quiet and level; and the equatorial band was somewhat more calm, more inclined to strati, and cirrostrati, than the tempestuous cumulostrati of the tropics. Judging too from the drawings, made solely with a reference on each night to putting in as much as I could of all that was visible, there appears no doubt of the medial line of calm, not being exactly coincident with the equator. Should this circumstance be borne out by future observations, it may be held to arise from the same causes which make the southern Trades overbalance the northern upon our earth, and throw the zone of so-called equatorial calm into north latitude—viz., the unequal distribution of land and sea surface in the two hemispheres. Such a result would be proving much, seeing that some theorists have been lately contending for Jupiter and all the outer planets being mere globes of water, with at most a cinder nucleus.

Since the publication, in a few words, by the Admiralty last October, of this discovery of cloud forms in Jupiter's belts,—M. Babinet has stated that a similar result had been arrived at in the Paris Observatory, with an object glass of nine inches aperture; and in this country Mr. Warren De La Rue has published an admirable plate of the appearance of Jupiter, as seen in his excellent reflecting equatorial, thirteen inches in diameter. His forms of clouds are by no means so clear, as they were shown by the seven and a quarter inch refractor on Alta Vista; but exhibit such an advance on everything that had been previously engraved, that I can only cordially wish that such an observer as Mr. De La Rue, with such a telescope as that reflector (it is of his own manufacture), may have an opportunity of trying its powers on a situation so favourable as Alta Vista.

There are two modes of decreasing the depth of pernicious atmosphere through which we are condemned more or less to observe all heavenly bodies; one is by elevating our station vertically, as already done at Teneriffe; the other is by, either biding our time, or shifting our geographical position, or choosing our object, so as to be looking in the direction of the zenith, more nearly than of the horizon. The advantage of so doing was exemplified in the greater revelations disclosed to us by Jupiter, than by Saturn; and was still further borne out by double stars. For, certain test objects of the northern hemisphere discovered by Struve, at Pulkowa, near his zenith—were difficult for us on Teneriffe: whereas the companion of Antares, a star near the horizon of the Russian, and never properly seen by him—was egregiously plain on the Peak, where it rose, from difference of latitude, 30° higher in the sky.

The 7th of September had now arrived, and we were beginning to fear that Don Martin Rodriguez had forgotten the invitation we had given him, when on Guajara in August, to visit us at Alta Vista in the first week of the next month, and see something through our large telescope, though its transport had been declared impossible.

We had just given him up, when there arrived two sturdy countrymen with a large skin of goat's-milk, a supply of white loaves and yellow fragrant cakes, a pasty of French partridges, and finally, the pleasing intelligence that the Don had been travelling all night, that he was at the moment resting amongst the rocks of Estancia de los Ingleses, and would presently complete the ascent.

CHAPTER V.

SUMMIT OF THE PEAK.

WITH the accession of Don Rodriguez, and his attendants to our party, an excursion to the summit of the Peak was organized. Accordingly on September 8th, two men being left at the station to observe barometer and thermometers every quarter of an hour, all the rest of us,—loaded with meteorological instruments, photographical apparatus and other knickknacks,—started off on foot by break of day.

A few yards beyond our walls, coming to the end of the pumice-stone ground, that furthest limit of horses and mules, we entered the wilderness of the " Malpays," or those ultimate lava streams of Rambleta, the torrents of black lava rocks and stones. At first the path led up a narrow angular valley, between the flanks of two adjacent ridges. The sides, inclined at a steep angle, were mere loose stones, of

which the larger blocks formed the bottom of the
channel, and had a little fine yellow pumice sifted in
amongst them. This was dust that had been carried
up by east winds from the plateau of Alta Vista, and
had not travelled far. Step by step the quantity of
it decreased, and before long we were walking up the
ascending angular passage, on nothing but black
disjointed stones.

Presently, large opposing masses, obliged us to
make a slight diversion. One of the sailors, who
was walking ahead; very zealous, and from having
visited the Peak a few days before, anxious to show
his knowledge,—went straight up the sloping side,
sending down instantly a shower of rolling stones;
and he would ultimately have arrived only at some
impracticable and overhanging shell-like surfaces of
lava. Manuel, the Spanish guide, smiled pleasantly
at the essay, went himself to the front, and in spite
of the immense load on his shoulders, paced lightly
from rock to rock, as with the skill of long experience, he led the way in a sinuous direction, and
enabled us easily to gain the summit or ridge-back of
the stream.

On we then went, stepping from stone to stone as
easily as on a staircase, for the blocks were generally

wedged in tight, and in size were seldom less than one foot, or more than three feet cube. Whenever larger ones occurred, and they did now and then appear, and of most fantastic shapes,—fragments apparently of some curving crust of lava, one-half dense and polished stone; the other half, cinder,— there was little trouble in turning their flanks.

Of this part of the Malpays, strangely different accounts have been given by able travellers; thus, Captain Glas, in 1761, leads us to imagine a flat sheet of rock cracked across into cubes. Humboldt, in 1798, says, "The road, which we were obliged to clear for ourselves across the Malpays, was extremely fatiguing:" and "the lava, broken into sharp pieces, leaves hollows, in which we risked falling up to our waists." That excellent observer, Von Buch, in 1815, mentions chiefly "the sharp edges of glassy obsidian, as dangerous as the blades of knives." And Dr. Wilde, in 1837, writes, of the scene, as "a magnified rough cast;" a simile which fails in conveying an idea of the utterly disjointed nature of the ground, composed of only loose blocks to an immense depth, and with no sort of cementing or filling up material between.

With such a formation, crevices of a certain size abounded everywhere. It would have been dangerous therefore to have dropped any small article, lest it should have fallen through from one crevice to another, until deep below the surface we were walking over. But the breadth of the gaps, though quite enough now and then to take in and break a horse's leg, is never, or by proper care need never be, anything inconvenient for even a lady to step across.

Where doctors differ, there is nothing like the testimony of a photograph; so we planted our camera, at a fair average part of the Malpays; and straightway obtained *Photo-stereograph, No.* 10; the handle of a geological hammer, on a stone in the foreground, presenting something of a scale for measure.

Slowly ascending, stopping every few minutes to observe the meteorological instruments, we arrived about half-past six, on the level of the ice-cavern, 11,050 feet high, and distant from us only about twenty yards. Nothing was to be seen externally, except the chaos of tossed and tumbled blocks of lava, somewhat larger than elsewhere; but not materially different from what was to be seen above, around, below, and on every side. Everywhere a wilderness of black stones, closed in our view, except when

looking down towards the east. 'Twere ungrateful
not to mention that, for had not the sun just risen
there, and had we not in truth admired?

We had done so, and carefully noted the changes
that occurred since early morning, from where at
first, amid nocturnal darkness, the only symptom of
approaching day, was the long glade of zodiacal
light, shooting upwards amongst the stars to Orion
and Taurus; and glowing towards the lower part of
its axis, so as to repudiate either the heliocentric ring
of one writer, or the geocentric ring of another.
Then after a while came the low flat arch of early
dawn, faint and blue. With humility it appeared on
the scene, and sat down in the lowest place; while
the zodiacal light aimed ambitiously at the highest.
Time passed on, and the proud one waxed faint, while
the lowly one was promoted to a higher and higher
position; unto it next was given a reddish hue, as a
dress of honour; and the lenticular form of the
zodiacal light was seen no more. A few minutes
further, and a yellow tint manifested itself in the
dawn, upon the red; extending below—it shed a rich
orange along the horizon; expanding above—it produced a somewhat cold, greyish, even greenish tint,
but one eminently luminous; and fit harbinger of

approaching day; lighting up earth, and sea, and the broad white clouds spread far and wide below us.

To the south-east, the volcanic peaks of Grand Canary rose in dark-coloured, angular battlements, through the sheet of vapour; but for which, in the E. and E.N.E. we might see something of the two lowest and most distant of the Canaries, Lancerote, and Forteventura; for their azimuthal direction lies open. Again a new illumination strikes out from the east; its yellow glow is intensified; and has almost overpowered the lower red; while the cold region of light at its upper limit, is now corrected by a magnificent blush of rose-pink, which stretches high up into the blue. Then the first point of the solar disc leaps up behind the horizon of an ocean of cloud, and darts his long vivifying rays athwart its cumulous masses.

Duly did we note, and intensely did we admire these brilliant effects; yet with justness it is hoped, as well as due regard to present and other times and places. Not so all previous authors, amongst whom appears to have obtained a dogma, that no account of an ascent of the Peak of Teneriffe, can claim to be legitimate, without a laudation of morning. Sun-rise is as necessary to this species of composition, as

to an epic poem; and why? The established mode
of making the ascent from Orotava, is to travel all
day, so as by evening to reach Estancia de los Ingleses. This spot, lying on the eastern slope of the
Peak, all view of sunset, and indeed of the whole
western sky, has been hid from the tired traveller
through the whole afternoon. But in the morning,
he wakes refreshed, and is extraordinarily excited by
finding himself for the first time in his life, two miles
above the sea. So when the sun springs forth from
the east, in that unclouded brilliance common to the
atmosphere's upper regions,—the traveller compares
it with dim sunsets far, far below, and declares sweep-
ingly in the words of Dr. Wilde—"I have often tried
to form to myself a comparison of sunrise and sunset,
and on this occasion have settled the question in
favour of the former."

Poor sunset is here treated very unfairly, and has
no advantage of elevation extended to her, though it
was measured out for her opposite, to the extent of
9700 feet. A preliminary residence on Guajara would
be an excellent corrective to the ideas of such one-
sided tourists; for there, the west horizon is as clear
as the east. I must add too, from a comparison of
all my sketches, made during my own noviciate there,

that the sunsets were on the whole more richly coloured than the sunrises. Besides this, the registers of radiation instruments erected there, show, during a whole month, that there is much heat given out by the western sky after the sun has gone down, but none by the eastern sky before he has risen; and hence an explanation of the comparative poverty of a morning appearance.

Pardon, pardon, O reader! this long digression. Had there not been such an elegant matter of opinion to discuss, how doubly weary, how utterly monotonous the way we are now treading.

By seven o'clock we have reached the height of 11,240 feet, very nearly the middle of this real Malpays. Far and wide it covers, or rather forms, the side of the mountain, with its loose black stones; observing a certain method too; for there is a grooving, and ridging, as different lava streams have poured tortuously down, like huge black serpents descending the Peak. One ridge is so like another, that an unskilled mountaineer might easily lose his way; rather however in the coming down than the going up; but at the chief turning-points, the guides have piled three or four stones one upon the other. To a

stranger not very noticeably; but after having lived
for some time in this wild Malpays world, where not
a precipice, not a flat, not a patch of smooth or soft
ground, not a plant, not a bird, nor even an insect
exists, where one's whole attention is taken up with
stones, stones, and nothing but stones, all of the same
black lava,—the eye becomes at last so nice in appre-
ciating small distinctions amongst stones, that the
three or four piled by man, become as instantly
distinguished amongst the acres of them piled by
nature, as if they had been an actual finger-post, let
in amidst the barbarous lava.

At 11,500 feet of elevation, we came to a sort of
chamber depression, and indulged in notions of what
an excellent observing-room it would form,—if we
could only get the telescope up so high,—protected
from west wind, without anything to produce dust
in its immediate neighbourhood, and with crevices
enough to swallow and make away with, any pow-
dery matter brought from a distance by a stray
whirlwind.

At 11,600 feet was a jet of steam, coming out of a
crevice or hole amongst the rocks, about three inches
in diameter. This was the well-known narix of the
Peak. The vapour, whose temperature was from 100

ALTA VISTA OBSERVATORY, FROM THE NORTHERN LAVA RIDGE.

v. 285.325.

Printed by A. J. Mullraish under the superintendence of James Glaisher, Esq. F. R. S.
and published by Lovell Reeve.

to 122° Fahrenheit, condensed on the neighbouring
stones, and gave means of support to a few handfuls
of moss growing between them.

This phenomenon passed; still lay before us the
stones, the black lava stones, as bare, but if possible
more rough, and rude, than ever. After a while, some
little sprinkling of pumice-stone dust began to manifest itself in crevices; more and more appeared; and
suddenly at the altitude of 11,745 feet, we emerged
from the Malpays. Instantly there rose before us,
high above our heads, the Piton or sugar-loaf cone,
forming the summit of Teneriffe, resplendent with
light red and yellow, like some huge tower, gleaming
in the brightness of the morning sun.

The place that we were on now, between the Malpays, and the Piton, was "Rambleta," by some
described as a plain; but there was so little flatness
about it, that we could not very easily get a convenient corner for our breakfast service. There is really
only a slight difference of slope between the Malpays,
Rambleta, and Sugar-loaf; while the space through
which the moderate angle of the middle locality lasts,
is inconsiderable. We were easily pleased, however,
and after contemplating with admiration, during our

meal, the high smooth slope of pumice, rising like the
cone of some giant glass-house,—we set off again;
and with all our apparatus on our backs, began the
last climb.

The mean angle of the sugar-loaf is 33°; on the
east, where all travellers ascend, it is about 470 feet
high; on the opposite side nearly 650 feet; its base,
Rambleta, dipping towards the west. At first, walking through the loose pumice, we rather wished for
the Malpays again; but coming to the projecting
points of some red lava crags, we found such very fair
footing, that the ascent ought not to be spoken of as
difficult by any man traveller. Here and there some
warmth was felt in holes and cracks of the rock.
The fissures increased continually in number and temperature; then a faint sulphurous smell was perceived.
A few hasty steps more,—and we were on the brim
of the culminating crater, in the midst of jets of
steam and sulphurous acid vapours.

Fagh!—on inhaling the first whiff one was inclined
to beat an instant retreat for a few steps; looking, for
the moment, with infinite disgust on the whole mountain, as nothing more than the chimney, 12,200 feet
high, of one of Nature's chemical manufactories. This
chimney, having been built at great expense, she was

resolved to turn it to account. We, curiously foolish creatures, had been innocently creeping up the sides; and were now astonished to find, on peering over the mouth of the long stalk, that noisome fumes were ascending from it.

Again we mounted up on the brim, and soon getting toned down to breathing mephitic exhalations, found the chief feature of the crater interior, some 300 feet in diameter, and 70 feet deep,—to be its extreme whiteness ; often white as snow, where not covered with sulphur. The breadth of rim was hardly sufficient to give standing room for two; so immediately, and in such a knife edge, did the slope of outside flank, meet that of inside wall. On the portion of circumference where we collected, the ground was hot, moist, dissolving into white clay, and full of apparent rat-holes. Out of these holes, however, it was, that acidulated vapours were every moment breaking forth; and on the stones where they struck, were producing a beautiful growth of needle-shaped crystals of sulphur, crossing and tangling with each other in the most brilliant confusion.

The north-eastern, northern, and north-western were the highest, whitest, and hottest parts of the crater walls. Towards the west and south they dipped considerably, and verged to an ordinary stone-colour inside; outside they were red and brown all the way round the circle. Hence it arose, that when in previous months we had looked from Guajara, some of the bleached interior surfaces of points on the northern brim, being seen through and over the southern depression, gave us the erroneous idea of a double crater; an exterior ring-wall of brown, and an inside one of white, material; errors of perspective, it now appeared.

Some short portions of the interior of the wall, are precipitous rock, 10 to 20 feet deep. But generally the structure has so crumbled away during long ages of volcanic idleness, that it is now, like a baron's castle of a long past feudal age, going to slow and certain ruin; falling downwards in a mass of rubbish, that tends to fill up the central hollow. All about the curving floor, my wife and Don Rodriguez wandered over the deep bed of fragments, searching for the finest specimens of sulphur; and, with the photographic camera, I walked through and through the crater more than a dozen times, in as many different directions to take the several views,—completely disproving thereby all alleged dangers of the "awful abyss," that one tourist described looking into with fear; after he had " crawled " up on the outside to a high pinnacle, from whence he could safely make the survey.

Only in the neighbourhood of the walls, is there much annoyance from puffing steam and vapour; while neither there, nor anywhere else, is more than a thin coating of sulphur, often bedewed with sulphuric acid, to be found. If all the sulphur on the Peak were to be gathered together, by scraping it off the stones, a long and tedious operation in itself, there would hardly be two barrows full obtained; and speculators therefore in England, need not incur the expense of sending up here to the height of 12,200 feet, for so scanty a supply.

Comparing his own observations with those of previous travellers, Humboldt concluded a cooling of this crater; Bertholet in 1830, in a similar manner concludes a heating, and speculates in a lively French manner on what a catastrophic destruction of men will ensue, when this hoary old volcano resumes its pristine energy. As far as we could make out, the ground is heated by the steam, or at least has less temperature than the steam which permeates it, and which indicated in the strongest vent holes only 150°; while the boiling point of water which we ascertained by careful experiment, in a deep cleft on the western side of the crater, is 191° 08. There would seem therefore to be no "high-pressure" at work, nor indeed any sensible difference in the effects on the whole, since the day of Captain Glas, nearly a century since.

The expiration of steam by the volcano, has rather a happy effect than otherwise; for tempering, as it does, the sharpness of an atmosphere of great elevation, it attracts a population of bees, flies, and spiders, as well as numerous swallows and linnets (*Fringilla Teydensis*). After the solitude and desolation of the arid and dusky Malpays, our sudden entrance into this bright, white caldron of the crater, with insects and birds flying about in numbers through the moistened air, seemed a new as well as a strange world. A remarkable little colony at least, an oasis of life and activity in the midst of an elevated desert of lava. During the few minutes that a previous visitor spent on this spot, he remarked the bodies of some dead bees, and jumped too hastily to the conclusion, of an "oblique current of air that brought them up to die." But the far greater length of time spent by our party on the summit, proved plainly, that the living bees, which swarmed there in such numbers were

perfectly at home; and if no food was to be found
for them immediately round about, was there not
Chajorra at a moderate distance, well clothed on
its southern flanks with *retamas,* whose abundant
white flowers are to bees so dear!

A magnificent feature of the whole volcano, is this
Chajorra. From the culminating point we looked
down on it over the western Malpays,—formed apparently of innumerable ridges of black lava stones,
similar to those we had seen so much of on the east,—
and then over a light-coloured flat, to where a dome-
shaped eminence arose. Through its pumice covering,
had been many special protrusions of red and viscous
lava, flowing similarly to those of Montaña Blanco.
Ages may have been spent in these operations, Chajorra looking all the time like that innocent eastern
dome. Then came a more violent period, during
which the convex top must have given way, broken
in, and so formed a huge pit. In this the liquid lava
rose, as in Kilauea, on Mouna Loa; and may have,
for long succeeding ages, shown an open boiling
caldron of the metal, nearly level with the edge.
Then the contents cooled, froze, and so filled the hole
across, with a level plain. After this had endured for

years, again the mountain experienced terrible convulsions, the filling material was broken up, and
ejected either in powder, or in re-melted streams of
lava, which we still see as they poured over the sides.

All was thus got rid of, save a portion which
adheres to the southern wall. There we saw it
adhering still; and there our photographs represent
it, with its level top contrasting with the rounded
sides of the mountain, and showing internal precipices, three times as profound, as any part of the
older *caldera* wall.

More than a quarter of a century ago, M. Cordier
lamented the neglect shown by travellers towards
this magnificent crater, three-quarters of a mile in
diameter, 10,000 feet high, and so totally unlike, says
he, everything that we know. His advice has only
yet been followed by one individual, and rich rewards
are reserved for future geologists, independent
enough to leave the beaten paths of ascent.

Interesting as is Chajorra viewed by itself, the
interest rises on comparing it with Rambleta and
Montaña Blanco; they are the triple heads of the
one central Peak or cone of eruption, that rises in the
midst of the great elevation-crater. Brethren are
they, with a more than *Siamesan* union, yet of different

ages; the oldest and the tallest being Rambleta, with its Piton crest. Montaña Blanco, the youngest born, we have described already, as a huge rounded tumulus of pumice and ashes of the red lava or obsidian period.

There might we leave it in the case of any ordinary geological formation; but with a volcano, a confessedly unextinct volcano like Teneriffe, with a vital principle as it were, and a power of growth within itself almost like that of an organic being, we have to take future increase into account. Hence, would we see what Montaña Blanco may one day become, we have only to look to Chajorra, and behold the Montaña with a terrible gulf on its summit, where the liquid lava rises and falls, with every pulse of the volcano's life.

Again, should we desire to see the future of Chajorra, which has passed through its Kilauea age of an open caldron, and is just beginning to pour a few black streams over its early exudations, let us look to Rambleta. There stream upon stream, and torrent upon torrent of black stony blocks and cinders have been forced forth unceasingly for ages. There too, a Piton has been formed in the centre of the crater of eruption, the whole of which has alternately been filled up by the pressing of material from below, combined with the fall of endless volleys of stones, shot up from the Piton, and rolling again down its sides; until now, the very *locale* of the Rambleta gulf is marked alone, by the black lava streams on its rim, cropping out of the rapilli slopes that lead to the final cone.

Tracing the circle by such markings round and round, Rambleta must have been a magnificent caldron, well worthy of the black lava inundations it has poured forth; and in the midst of it, the "sugar-loaf," with its little terminal crater,—or the present culminating point of the mountain,—must have evidently played only the same diminutive part, that does the small cone inside the crater of Vesuvius.

This cone of Vesuvius, as described by travellers, keeps ever shooting up stones, when the grand boiling and seething of the fiery lava, for the destruction of towns and villages, is going on round about its base, in the wider crater; over whose edge the matter presently pours, seaming the sides of the greater cone, the true Vesuvius, with a black lava ridge; not very different from the spectacle of Alta Vista.

So likewise the terminal sugar-loaf of Teneriffe appears, in its day, merely to have shot up stones and pumice ; no lava stream having ever exuded from its mouth. This privilege was restricted to Rambleta, the analogue of the Vesuvian crater ; as the whole Peak from the floor of the great crater of elevation, is the *Vesuvius ;* and Guajara, the *Somma,* of the Island. When therefore, the usually sagacious Humboldt, comparing Teneriffe and Vesuvius (" Personal Narrative," vol. i.), contrasts against the latter, merely the sugar-loaf of the former ; and so finds for the proportion of the " ash-cone of Vesuvius" to the entire mountain $\frac{1}{3}$, and for that of Teneriffe only $\frac{1}{22}$, —he infers a needless anomaly, somewhere between $\frac{1}{2}$ and $\frac{1}{3}$ being the true proportion for the latter volcano.

Much has been written too, at various times, on the supposed law of decrease in the diameter of craters with the increase of their height ; such position being defended by comparing the breadth of the small terminal crater of Teneriffe with the full bore of Vesuvius, combined with their absolute elevations. The facts compared, however, not being analogous, results thence arrived at, fall to the ground ; and if we again make a comparison, keeping strictly in view

the principle of like with like,—against the bore of
Vesuvius and its height above the plain of Somma
on one side, setting the united areas of the caldrons
of Rambleta and Chajorra, with their elevation above
the Canadas on the other,—we shall find both
diameter and height equally magnified; Vesuvian
dimensions being in every way quadrupled on the
Peak.

Similarly also comparing Somma with Guajara, we
shall arrive at equally favourable results for the
latter; and find in fact, that in all its craters,
Teneriffe is of standard proportions, as well as of
gigantic size. That size is vast, but evidently,
nevertheless, gives no indications of approaching
near the limits, of nature's volcano-making power.

With all this breadth and height then, what of
the view? The view has been so often described
by able writers, that I will merely mention a green
band around the middle and lower part of the mountain, produced by heaths and grasses; a large expanse
of fir-trees (*Pinus Teydensis*), on the north-western
slope; an eruption crater there, as red as fire, others
about the coast line, especially towards the south,
brown and black, and gaping open with wide

mouths, like the fish of "Arabian Nights'" story, standing on their tails, and answering the summons of the magician. The two near Capes or corners of the island, Teno and Rasca, and the long far off one, Anaga; altogether, rivalling the triangular figure of Sicily—Mediterranean home of Etna.

Beyond is the sea, a small portion of it only is visible; for soon beyond its shore, begins the edge of the cloud-level; and it is that, stretching away over the waters to the distant horizon, which there joins the sky, and seems to support its vault of blue. The tops of the islands, Palma and Grand Canary, are seen above these lower clouds; Gomera and Hiero, where no cloud is; but for Lancerote and Forte-ventura we look in vain; they are safely covered in beneath the cloud that floats at twice their height, and overspreads all the sea in their direction for hundreds of miles. The Don has heard that these two islands are visible from the Peak, but has never seen them himself, and cannot see them now; nor can we; nor did we ever do so during all the mornings we spent on Alta Vista and Guajara; so constant throughout the summer, is the dense, broad plain of N.E. cloud.

Enough of the view, the west-wind is blowing

strong and cold; and we shall do well to descend, and
get through the Malpays, before the mountain-night
arrives with its sudden darkness. Yet wait my
friends, said I, one moment more; allow me only
another photograph; for these sulphurous exhalations
of the ground, have spoilt nearly all my plates to-day.
So Don Martin Rodriguez placed himself again on
the highest point of the crater-wall, absolutely the
culminating point of the Peak of Teneriffe, his man
stood close by, and the yacht carpenter going past at
the moment with a bucket of sulphur specimens, was
included in the picture (*Photo-stereograph, No.* 1);
where the dark rocks in the foreground, show the
brown exterior,—and the white cliff under the Don,
the acid and steam-bleached interior,—of the terminal
crater of Teneriffe.

CHAPTER VI.

AUTUMN IN EXCELSIS.

AFTER our visit to the caldera of the Peak, with its walls of rock bleached by steam, and by acid vapours permeating them for ages, we could better understand the remarkable internal whiteness of lunar volcanoes, as shown by our telescope at Alta Vista.

Some geologists have, indeed, denied that the features seen by astronomers in the moon are to be considered as volcanoes; but we who duly noted the gentle external slope of some of those circular pits, their cliffy internal descents, their flat floors, and their central peaks—had little doubt in our minds. Occasionally could be traced something much like a collection of stony lava streams; which even the Spanish attendants, when looking by permission into the telescope, would call a Malpays. Generally too would they describe what they saw, with the same terms that they employed, for volcanic features of the mountain whereon we stood.

Could we have found in the moon, that dynamic trace, which was so important in proving relative ages among the red and yellow lavas of Teneriffe, viz., glacier wrinkles in one, and surf-like waves in the other,—all sceptical doubts must vanish. But we failed, and this point is left to a larger telescope, more constantly employed in lunar physics, on this or some higher mountain.

Details of a larger sort, however, were multitu-dinously brought out by the Pattinson Equatorial; to such an extent, indeed, as to be hopelessly beyond my poor efforts, to record them usefully in the small portion of time, that was available each evening. For it is to be remembered, that there were high lava ridges westward, and the young moon had very low south declination. Night after night, this latter quality lessened; and so improved; but then, the planet drawing near her full, became less appropriately illumined to show configuration of surface.

Our lunar observing season therefore, though rich in promise, was very short, and by no means equal either in quality or extent to what, with a little difference in the preliminaries, it might easily have

attained to. This was not our only misfortune; for
with the general brightening of the sky, consequent
on growing lunar illumination, the nights became
less appropriate to trials on small stars, for space
penetrating power; and thus, lamentable barrenness
began to follow close, on our recent nights of unexampled astronomical harvest.

We complained, however, the less, inasmuch as on
each succeeding evening, west winds blew stronger
and stronger down upon us from the Peak. By day
there was an east wind blowing up, from the crater-
plain, with a velocity of five or six miles an hour; but
at night it was made up for by increased force in the
western squalls. These came at last so fast and frequent, from sunset to sunrise, that the telescope shook
to a degree, rendering it impossible to take accurate
observations of any sort of objects, bright or faint.

September 11th was another term day, but its
promises were of a negative kind. Hour by hour we
saw the barometer falling, and before six o'clock of
next morning, the S.W. cloud had arrayed its forces
once more over land and sea. The north-eastern
stratum, confounded at this resuscitation of its
enemy, seemed struck with panic; and immediately

broke up into detached portions, ready for flight.
Still the solar radiation was strong, and with all our
experiments to try to make manifest the " eclipse red
prominences," the telescope eye-pieces got danger-
ously heated ; and in spite of care, some part or other
of our photographical apparatus, for picturing the
sun's image, would every now and then begin to
smoke and burn.

The bad symptoms, predictions of future, rather
than manifestations of present, evil, nevertheless
went on increasing all the time. September 12th
was cloudy, and though the sky cleared up again the
following day, the 14th opened with bad omens. A
whistling S. wind; and, most unusual feature, on
the sea between Teneriffe and Grand Canary, long
curving lines, indicating a grand current of sea, or of
wind acting on the sea, from the S.W. These current forms were traceable, in their curving inlet-
sort of markings, for a distance, each of them, of
twenty miles unbroken. Earnestly did we look at
them, thinking what they might betoken.

Yet still the sun was shining; the barometer too
had risen a full tenth in the last day and a half; and
in the course of the forenoon several parties of visitors
arrived. The news of what Don Rodriguez had

ENTRANCE TO THE ICE CAVERN IN THE MALPAYS ON THE PEAK OF TENERIFFE.
AT THE HEIGHT OF 11,040 FEET.

_p. 361 363._

Printed by A. J. McDhwaibrander the superintendence of James Glaisher, Esq.F.R.S.
and published by Lovell Reeve.

seen through the large telescope, had spread far and wide; and others now hoped to enjoy a similar opportunity of star-gazing made easy.

This was all well enough: but they brought us a letter from the captain of the yacht in Santa Cruz roads, relating how, when he was quietly one morning at the moorings where he had been lying for two months, a large man-of-war that had anchored between him and the shore three days previously, had put to sea in such a clumsy fashion, as to run right upon him. Had the yacht been wooden-built, she must have been cut, he said, down to the water's edge; as it was, the strength of her hull had been sorely tried, and a large part of her bulwarks carried away. To repair this damage, he now earnestly requested us to send him the carpenter.

We were all busily engaged in discussing the strange, the inexplicable contretemps, when suddenly the sky darkened, and down came a rattling fall of hail, that soon whitened all the ground. The painted canvas was instantly brought out, and every part of the costly telescope and mounting, safely covered up.

The hail ceased, and then there was a sudden agitation amongst the guides, who had accompanied our bands of visitors. These men declared to their

employers, that such intense cold was coming on, as
would kill their horses, should they stay there
through the night; and that all must descend instantly. This was not agreeable to the tired travellers, and they appealed to me.

The barometer had passed its lowest point, therefore the rain, if coming, was at hand. The mercury
was oscillating violently, a sign of much wind. And
finally the difference of wet and dry bulb thermometers, reduced to one-fourth of its usual amount,
was too sure a symptom that rain was intended, and
in no measured quantity.

Immediate preparations were then made by the
visitors, to conform to their guides' advice. Bottles
and pies and loaves of bread were bundled again into
saddle-bags, big boots were pulled on, great-coats
and mufflers mounted once more. The guides never
worked so actively, as they did now in saddling and
loading the horses. One after another, each went off
as soon as he was ready.

The last passenger was Herr Kreitz, to whose
mechanical skill I had been so much indebted for the
successful re-arrangement of the telescope in Orotava.
He was exceedingly desirous of looking through the
instrument, and was utterly exhausted by his long

ride performed through the night, and by a sick
headache. His horse was saddled, and standing with
his head turned towards the descending slope, below
whose edge his companions had already disappeared.
He had bidden adieu, and had departed five steps,
when he returned with, "If the clouds should dis-
appear at night, will you uncover the telescope and
show me the moon?" There is not the smallest
chance, said I, of the clouds disappearing; and Herr
Kreitz went griefful away. We were sorry for him,
for he was a man with a soul to appreciate nature,
as well as a quick brain and ready hand for exact
engineering.

All the bustle and confusion of the morning, with
the unexpected arrival of so many horses and mules,
English and Spanish visitors, Orotava men and inhabitants of Santa Cruz,—had now died away, and
we were a very small party left behind, to stand the
brunt of whatever might be coming. Before it
actually came, however, our number was destined to
be still further reduced. Our servant Manuel had a
dog, one "Pēcho," a brindled animal with ears like a
fox. When first brought to Guajara, he was a lean
anatomy of skin and bone, fierce as a wolf, and sus-
picious as a hyena. Before very long he found out

the state of society in our encampment; and while
preserving his savageness to strangers, became very
gracious to my wife and self; and got quickly fat-
tened up in consequence. This dog Pēcho then, came
out on the present occasion, wise as a serpent and
fully as sleek, took one look at the sky—and concluded that he had better go down with the horses.

Darkness advanced with unexampled rapidity; the
violence of the wind, and the moisture of the air
rising in proportion. At 4 P.M. there was a heavy
cloud on the mountain-top, descending to the level of
the ice-cavern. At 4h. 30m. masses of vapour were
seen rolling over the great crater below us, and at
5 o'clock we ourselves were enveloped in dense mist;
unable to see what was coming next, and compelled
to abide whatever it might be.

The meteorological instruments were now our only
means of questioning the weather, and very sorry
comforters they were. The oscillating barometer
confirmed itself by the continued increase of the gale,
blowing due from the south; while the dew-point,
only 2° below the temperature, when at 9 o'clock A.M.
it had been as much as 40°, prepared us for abundance
of wet.

Forewarned, was not altogether forearmed, and thereby defended; for not until the rain came on, as it did about 7 o'clock in the evening, driven horizontally by the furious south wind, did we know all the various points of weakness in our habitations. The roofs, mere lengths of felt laid over sticks, were too flat; they bagged in between their supports, and began to leak at every yard. Then the felt would be lifted up by the wind at some corner or joint; a dangerous flapping would begin, and unless immediately stopped,—by our rushing out in the fog, rain, and darkness, finding a heavy stone, flattish if possible, and clapping it on the peccant morsel,—would threaten to carry away the whole roof.

Meanwhile the barometer showed a small inclination to rise, but the dew-point depression decreased to 1°, the temperature being 40° 5. The wind by midnight rather deviated from its early, steady, unmitigated force, and had become somewhat squally; but all the gusts, strong and weak, came equally from one and the same direction all night through, viz., due south. The high black lava ridge there (*see Photo-stereograph, No.* 12), that we had looked to, for defending us from such storms, was not only perfectly invisible; but one would have believed it quite removed, such a clear, clean blow of it did the south wind make upon us.

The rain kept on without cessation. After months of dry weather, nature seemed bent on balancing her accounts. For a long while we made ourselves quite easy as to ultimate consequences, for could not the thirsty soil swallow up and dispose of an almost infinite quantity of water, without inconvenience? But as hour after hour of this heavy liquid-fall went on, and as buckets and basins set to catch droppings from the roof, were filled and emptied, and filled again with alarming rapidity, a new found fear began to arise in our minds!

Might our walls be at last undermined? Was this low tract of pumice on which we were encamped, between the lava ridges on either hand, the bed of a winter torrent? Might we presently expect to hear the rushing sound of many waters above us, as the cascade broke into life, and came rushing down the steep slope upon our devoted walls, that stood exactly in its pathway? We could not assure ourselves of a negative. Practice and experience alone could settle the question.

All that we knew was, that the walls were in a very hazardous condition. Dry walls, run up in five days by a few men, what could be expected from them? Especially when remembering the bad shape of the stones! No more flat slabs as in the laminated, half-stratified, submarine-formed trachytes of Guajara, were here; at Alta Vista were only the rolled and fallen lumps of sub-aerial lava.

Had it depended on the shape of the stones alone, our walls would long since have fallen. But happily, they had a roughness peculiar to themselves; a roughness which, while distressingly damaging even to the horny hands of the builders, had the admirable quality of making stone stick to stone, almost like a couple of brushes driven into each other, bristles against bristles.

Thus then did the materials of our walls cling together, almost in defiance of gravity; and so different did their minute vesicular nature make them, from ordinary stone, that the carpenter found he could drive his nails into them with the greatest ease; and he had so fastened, a sort of tapestry around the walls of our room.

How those cloths did wave and flap about that night; for the walls behind them were porous, very porous, and were becoming more so every instant. This unpleasant feature arose from the expeditious method that had been adopted in the building; each wall being formed of two built-up faces, with their internal void hastily filled up with sand and rubbish.

Excellently did this plan answer for a short time; but each succeeding day, as winds continued to blow, and walls to shake,—the smaller of the filling-up particles were gradually falling downwards, and leaving the interstitial division above, quite empty. Not only so, but by their undue pressure on the middle, they were always tending to burst out either one or other face; and now the rain entering, assisted the inside material, in its downward flow, and outward pressure.

Had we not most fortunately, employed ourselves during many previous evenings, in wedging up all the external crevices of the walls, giving them thus an increased strength, it might have fared badly with us, in this long, long night, of stormy wind and driving rain.

With morning, the rain happily ceased; the mist withdrew a few hundred feet above our heads, the wind moderated in strength, and I went out to survey the mischief that had been done.

Nearly two inches in depth of water had fallen in

the course of the night. Sad to say, the whole of
what had descended on the sailors' roof, had gone
through the middle of it, right upon them as they lay
on the floor below. Yet withal they never looked
more cheerful, than when I hailed them at daybreak
through their broken doorway. They had been
nearly wrecked they said, but smiled at the same
time, taking it quite kindly, as if it were something
they were perfectly accustomed to.

They did also, in truth, physiologically prefer the
present temperature of 40°, saturation of moisture
and a clouded sky, to the dry heat and vivid radiation of previous months. Then, they were dried up,
oppressed, discouraged; while the Spanish Manuel,
with his brown complexion, and wiry frame, would
ascend or descend like a chamois amongst the rocks,
in the blinding light of a vertical sun. In all this
arid heat, he would come down from the ice-cavern
at a run, bounding from block to block of the
Malpays, with *two* water-barrels on his back, an
example to all beholders.

Now, what a change! Poor Manuel looked like a
drowned rat, and could only shiver and shake, and
whimper that he must go down. The sailors, in their
glorious strength of Saxon constitution, generously

excused him ; he was not sufficiently clothed, they
said, only a cotton shirt, linen drawers, a straw hat
and raw hide shoes ! What cold could he withstand?
Out of their slender stock they fitted him with a
Royal Yacht Squadron guernsey, and several woollen
articles. But he continued so wretched and que-
rulous, that at last we gave him leave to descend the
mountain, but with orders to be up again within three
days. No sooner said, than instantly he disappeared,
as quickly as his dog Pēcho had done the previous
evening.

A nice little party were we now at the station, for
scientific work, had the weather been favourable; my
wife and I in our house on one side of the telescope
square, and the two yachtsmen in theirs on the
other. We four were, for the time, in possession of
the whole summit of the mountain, of full half of the
Peak of Teneriffe; in so far as there was not a single
human being within many miles to dispute our
claim.

The sailors were lively and ready to assist in anything. We observed meteorologically, cracked the
biggest blocks for fresh fractures; dug into the
pumice of antepenultimate lava streams to test their

depth and temperature; but nothing astronomical did the sky and clouds allow; the season had effectually changed. The gale of August 30th-31st had blown the note of warning,—the present storm had accomplished the fact,—for summer was gone, and autumn verily begun.

Everything about us altered utterly. No more *lower* clouds now, the chief feature of the Trade-wind; that must have perished also. In place of this current, a continual wind blew on us from the south, day after day; having its cloud stratum far above our heads. We were living now below the clouds, like men on the sea level; as astronomers do in most of the public observatories. Hence no longer to us were clear blue skies, and bright luminaries; but the sun by day and the full moon by night, if seen at all, were faint and hazy, scarcely struggling through diffuse and watery clouds.

Watery too was the air round about us, and in the place where our best seasoned mahogany boxes had been cracking and rending with the drought in previous weeks, we were now unable to get our bedding moderately dry.

The temperature still kept to 40°, and the dew-point depression was a mere nothing. Occasionally

a shower or two of rain would fall, sometimes a
sprinkling of hail.   Yet withal was the weather
bland and genial to our feelings. The peculiar sharpness of mountain weather, with its opposite radiations
by day and by night when the sky is clear,—had
gone; and with it, the astronomical advantages of
the site had ceased and determined for the season
of '56.

CHAPTER VII.

THE REITERATED QUESTION.

SOME three days after the storm, when our reduced party was engaged in packing up the telescope, —unexpected voices broke the stillness of the mountain air. Above the edge of the slope leading down to Estancia de los Ingleses, suddenly rose a head, and then the shoulders, of a stalwart Spanish peasant. His whole figure manifested itself quickly as he advanced, walking rapidly towards us. Behind him, other heads and other figures similarly rose up from the lower ground, with knapsacks on their backs, and iron-spiked poles in their hands, following their leader. When he had approached within a few paces, he doffed his hat, and pronounced the name of our friend, Don Martin Rodriguez.

The Don had sent a letter, and as usual, his men were not empty-handed. We rejoiced once more in his rich goats'-milk; as luscious as cream, and capable of being preserved in this high locality to the

eleventh day, by boiling with a little sugar and
water, and pouring into glass bottles. Then they produced a basket of fine fresh eggs, a luxury unknown
for months, and a new goats'-milk cheese. We felt
ourselves fortified once again, for another storm.

The leader of this party was a fine open counte-
nanced fellow, a genuine specimen of the worth and
strength of the country peasant. Our communi-
cations were not very fluent, from his understanding
no other than his mother tongue, and our not having
picked up much Castilian; but he was voluble in his
inquiries as to how things had gone with us during
the wet weather.

Quite regardless of rest or cold, had our present
letter party, come across the country from a high
level on the south-eastern flank of the great crater.
A tremendous walk! And now they sat down in a
corner to discuss their simple food, a little gofio,
while I answered the Don's epistle. The response to
our well-wishing friend on his own score, was soon
indited; but then he had enclosed a note from a person or persons unknown, begging and praying of an
answer to certain scientific queries, astronomical and
meteorological, drawn out to the extent of two closely
written pages.

The paper was dated, Orotava. Who could be the

author? The Don said a friend of his; so we stretched out to the task of answering. How much we were interested some few days after, to learn, that the unknown was a grey-headed Spanish gentleman, who for twenty years past had been an utter invalid, and had sought for rest and relaxation in scientific reading. For the sake of its books, he had taught himself our English language, and had profited by its literature extensively.

The very first of this gentleman's questions showed an able mind, for he had picked out from the third volume of Humboldt's "Cosmos," precisely the problem, which, from my peculiar situation, I should labour to answer, and which my official instructions had ordered me to undertake. The question was one, towards whose solution, the author of that famous book, had vainly, through fifty years, called for additional observations to his own; and he takes that opportunity pp. 55-56 to reiterate his request; concluding with, "as the Peak of Teneriffe is so near us, and is so often visited before sunrise by scientific travellers provided with instruments, I may hope that my renewed request for the observation of the lateral fluctuations of stars may not be without effect."

The phenomenon seen by the learned Baron was

sufficiently extraordinary in itself, and its knotty
points were but twisted more inextricably, when called
"lateral refraction." Generally, the idea of such a
refraction, in the atmosphere at large, is "unsound;"
yet if after having eliminated exceptional local influences, we restrict ourselves to exceedingly minute
portions of time and space, as seconds and fractions
of seconds, or quantities visible only in a large telescope with a high magnifying power,—it may be perfectly sound and true.   The circumstances however of
the case under discussion, denoted irregular movements of the stars, by no means confined within telescopic limits; and were indeed so wide and discursive,
as to arrest attention, and excite astonishment,
amongst the guides, as well as the savants, present on
the occasion.

To all this must be added, that the anomalies
were seen on their gigantic scale, precisely in a
locality where they should have been most liliputian,
or even microscopic, viz., in the thin, breezy air of a
mountain top, high above the vapours and emanations,
radiations and dust, and in fact all the disturbing
sources of both plains and sea. Theory therefore seemed
to be at fault. Where men expected that the motions
of the stars should be most regular, there they had

been found most irregular; or, contrary to what was computed and predicted for them.

It is precisely by attending to such, at first very unacceptable results of observation, and following them up by all the varied means in their power, that astronomers have from time to time succeeded in discovering an unseen planet, or arriving practically at a better knowledge of the varied consequences of gravitation, on different members of the solar system. So, if for a moment, some of them were disappointed at there having been found by observation one morning on the Peak of Teneriffe, traces of some general law of nature, new to theory,—they soon recovered the proper philosophical spirit; accorded abundant praise to the watchful traveller; and,—unable to co-ordinate what he had seen with anything that they were acquainted with,—called for additional observations.

Hence when one of our first philosophers was informed by the Admiralty of the intended starting of the present Teneriffe mission, he advised officially, " that the observer's attention should be directed to any instances of lateral refraction, like that remarkable case described by Humboldt as having occurred to him."

z

Similarly the Royal Astronomical Society requested the Admiralty, " that observations should be made to verify Humboldt's remark, on the lateral oscillations of stars near the horizon, and on scintillations in general."

Some private friends also brought the passage from Humboldt's work to my notice, before I left England. One of them, most sympathizingly deplored the indications thus presented by the mountain, of being totally unfit for astronomical purposes. Another hoped, that in a region where such effects culminated so ostensibly to their maxima,—the obscure and difficult subject of lateral refraction, would be looked into, with a good chance of being settled at last.

Loaded thus with the instructions and suggestions, forebodings and hopes of all parties,—we sailed from England. Time passed, and we were at length encamped within a few yards of the identical spot from which the Baron had witnessed the abnormal phenomena; and while still there, came this anxious letter of the unknown of Orotava, endorsed by our friend Don Rodriguez, asking if any explanation of the " lateral oscillations of stars," had yet been arrived at.

His postmen were unfortunately limited in time, for they had a very long and arduous walk to per-

Photo-Stereograph 14

EUPHORBIA CANARIENSIS ON THE SEA-COAST OF OROTAVA.

p. 406.

Printed by A. J. McGlashin, under the superintendence of James Glaisher, Esq., F.R.S.
and published by Lovells Reeve.

form before dark; I, on the other hand, could not express the result of my inquiries in a few words. It was not the case, such as often occurs in astronomy, of a few figures; were they right, or were they wrong,—it was rather the general and combined meaning of a page or two of eloquent writing, that had to be examined into. A description in words,—printed and published by its gifted author in several languages, and with some variations, during a period of nearly half a century—contained the scientific difficulty.

The answer so earnestly requested was, with some subsequent additions, much as follows.

For the facts we may refer to Humboldt's " *Relation Historique,*" in its original French edition of 1814, but for convenience on this occasion it is better to employ Bohn's English edition, 1852, which is a remarkably faithful translation. The proceedings of our traveller are perfectly clear; he arrived in Santa Cruz roads on June 19th, 1799, and left them six days afterwards, or on June 25th. In the interval, the 20th was spent in crossing the island from Santa Cruz to Orotava; the 21st and 22d in ascending and descending the Peak; the 23rd in resting at Orotava;

and the 24th in returning to Santa Cruz. This was Humboldt's first and last visit to Teneriffe.

The ascent began on the morning of the 21st, and seems to have been managed precisely as all the ascents of passengers from ships, and of hasty visitors, are still conducted. They start from Orotava with mules and horses early in the morning, and riding all the way, reach Estancia de los Ingleses, at an altitude of 9700 feet, the first night. Before daybreak next morning, they set out on foot; climb to Alta Vista, at 10,702 feet; pass the Ice-cavern, by one invariable pathway, at 11,050 feet; the Narix, at 11,600 feet; mount the cone, 12,200 feet, stay half an hour, and then hasten down to breakfast at the Estancia. There they take horse again; and descending quickly, reach Orotava that same evening.

Such appears to have been M. Humboldt's proceeding; he spent the night of the 21st at the Estancia (pp. 66 and 67), and about three next morning started on foot with guides and torches, reaching the site of our station, Alta Vista, "after two hours' toil." Whence he arrived, after "climbing over the broken lavas of the Malpays," at the Icecavern, a place which we visited daily for a month, to procure our supply of water. At this cavern it was, that the remarkable case of lateral refraction was seen, so that the identical spot where the Baron stood, was exactly ascertained, and frequently trod by our party.

The case itself he thus describes, (pp. 69 and 70): " While we were climbing over the broken lavas of the Malpays, we perceived a very curious optical phenomenon, which lasted eight minutes. We thought we saw on the east side small rockets thrown into the air. Luminous points, about 7° or 8° above the horizon, appeared first to move in a vertical direction; but their motion was gradually changed into a horizontal oscillation. Our fellow-travellers, our guides even, were astonished at this phenomenon, without our having made any remark on it to them. We thought, at first sight, that these luminous points, which floated in the air, indicated some new eruption of the great volcano of Lancerota; for we recollected that Bouguer and La Condamine, in scaling the volcano of Pichincha, were witnesses of the eruption of Cotopaxi. But the illusion soon ceased, and we found that the luminous points were the images of several stars magnified by the vapours. These images remained motionless at intervals, they then seemed to rise perpendicularly, descended sideways, and re-

turned to the point whence they had departed. This
motion lasted one or two seconds. Though we had
no exact means of measuring the extent of the lateral
shifting, we did not the less distinctly observe the
path of the luminous point. It did not appear double
from an effect of mirage, and left no trace of light
behind. Bringing, with the telescope of a small
sextant by Troughton, the stars into contact with
the lofty summit of a mountain in Lancerota, I observed that the oscillation was constantly directed
towards the same point, that is to say, towards that
part of the horizon where the disc of the sun was to
appear; and that, making allowance for the motion
of the star in its declination, the image returned
always to the same place. These appearances of
lateral refraction ceased long before daylight rendered
the stars quite invisible. I have faithfully related
what we saw during the twilight, without undertaking to explain this extraordinary phenomenon, of
which I published an account in Baron Zach's Astronomical Journal, twelve years ago."

This completes the account of the observation, in
whose accuracy and sufficiency its author does not
appear to have had the slightest doubt. What was

the cause then of so strange and even violent a phenomenon? Was it, as he goes on to hint, some effect
of the approach of the rising sun disturbing the
various layers of the atmosphere, and so producing
abnormal refraction? If so, it should be frequently;
if not always, seen; and over the whole mountain, if
not over most part of the world. Humboldt never
had an opportunity of looking again from the Peak
of Teneriffe; but from the South American mountains he tried the observation on one occasion; and
in the low country there, often; yet always without
success.

The immediate result from this, can hardly be anything else, than that the cause of the Teneriffe phenomenon was not anything of a very general nature.
But let us proceed in the inquiry.

My own experience of two months' observing on
Guajara and Alta Vista, the latter very close to the
" Ice-cavern" site, ought to have included cases of
the phenomenon, if of a wide-spread character.
But never on any one occasion was anything of the
sort observed. Yet the eastern horizon was an object
of frequent attention before sunrise; first to the naked
eye, on account of the zodiacal light, and the duration
of twilight; and next, to the telescope, on account of

Saturn; an object which was frequently scrutinized in the Pattinson Equatorial with a power of 500, and with clock motion applied. Under such circumstances, and with such means, a very small approach to the deviations described by Humboldt, would have been seen indeed; for the planet would have been thereby thrown altogether out of the field of view. But so far from that taking place, I have never in any observatory, either at the Cape of Good Hope, or in Great Britain, seen the planet so steady, and so well defined; though looked at there in parts of the sky much better adapted to producing good definition, than the low eastern direction in which it was only visible during our Teneriffe experiences.

These modern results then, would seem to coincide with Baron Humboldt's American observations, and to militate with them, against the idea of his phenomenon of June 22nd, 1799, being of a necessary character, as depending on the broad relations of sun and atmosphere. It must have been indeed entirely exceptional, even on the mountain where it was observed.

Of all the other visitors who have been there, not one is stated to have seen anything similar, except Prince Adalbert of Prussia, in 1847. His case is mentioned by Humboldt in the third volume of "Cosmos," pp. 55 and 56, as having occurred at the same hour of the day as his own, and at the same place, viz., the ice-cavern.

This fixation of the very ice-cavern as the spot from whence the star disturbance has only been seen, is not a little significant; for within a few yards of that locality there is a powerful agent in producing to an eye, in a given position, fluctuations in the appearance of stars, viz., no less than a vent of hot volcanic breath. This one cause is quite powerful enough for all the effects observed; and there are many such centres of optical disturbance about.

Thus, a little higher up in the same Malpays, is the well-known Narix, throwing out puffs of steam, and so much hot air, that our barometer observations were entirely vitiated at that point, in a certain hypsometric journey which we made : and the state of the air was rendered so exceptional, that the usual barometric formula indicated a hump, 300 feet high, on the mountain-side, at a place where there was really a continuous slope.

On the terminal crater again, are a thousand such vents. They are not always at work; and their intensity of variation is such, that it might be well

worthy of some future mission to investigate their
habitudes during a long period.   On some days we
saw white wreaths momentarily jetting upwards,
in dozens together ; by the eye—though not by the
ear, for there was no noise—leading one to think that
the mountain-top, was covered with sharp-shooting
soldiery.   On other days nothing was visible at a
distance ; and close by, only a barely perceptible
breath of steam exuded here and there.   But the
ground about those places was constantly hot ; and
though over a small extent, yet in their immediate
neighbourhood, always producing more or less refrac-
tive disturbance.

Unless then, such a direct and peculiar local cause
has been first eliminated from any man's observations, it is vain to occupy oneself in trying to explain,
what he saw at the place, on astronomical grounds :
vainer still, if there are manifest inconsistencies in
the original statement.   Now such there unfortunately are, by whatever accident produced, in the
immortal " Relation Historique."

The matters in question may perhaps be considered
by some, as trifling and allowable errors to have been
made by a young traveller, hastily ascending a mountain in the dark, for the first time in his life.   But

as I had been officially ordered to undertake the inquiry, I could not do otherwise than examine the whole account on its absolute and intrinsic merits. And having so become painfully aware of the existence of points, which would either prevent any hypothesis being founded on the narrative at all, or lead to erroneous ones,—it cannot but be anxiously desired, for the sake of science, that such questionable features should be examined, and rectified if possible. This is all the more important, as the illustrious reputation which M. Humboldt has since so deservedly acquired, imbues every word from his pen, with extraordinary importance in the eyes of all nations; amongst whom the particulars of his " fluctuations in the position of the stars," is still being repeated in successive editions, of the most famous book of travel, that has ever yet been published.

CHAPTER VIII.

THE ICE-CAVERN.

ON the morning of September 17, a file of men
and horses having arrived from Orotava, the
telescope, repacked in its many boxes, was sent down
the mountain side under charge of the sailors. After
duly depositing their burden with Mr. Goodall, they
were to rejoin the yacht, and assist in repairing the
unlucky damage that had been reported by the
captain.

Their tales of the ups and downs they had been
through, we were afterwards informed, were almost
endless; and for three days they talked continuously.
If they spoke of their doings as well as their sayings,
they must have described much excellent work that
they had both performed, and in the best spirit;
though the circumstances had not been nautical; and
the position, two miles above the sea level, not a
usual one for sailors. At the time that Mr. Stephenson so kindly allowed us to take his men up the Peak,

I had no idea of how much use they would become; and even he, perhaps, did not know all the varieties of labour that they were capable of; for the genius of the place, and the necessities of a strange style of life, were required to develope them. As the worthy fellows left us that morning, I could not but shake them both heartily by the hand, and thank them for their co-operation; wishing them well, as I still do, in their climbing the mountain of life.

As soon as the bustle of departure had passed away, my wife and I prepared for a photographical excursion to the ice-cavern, accompanied by a Teneriffe boy who had been left behind by the chief guide to wait on us. We three, were the only human beings left on the Peak, and the silence of the desert returned once more. The day was calm, cloudy too; and though so much rain had fallen lately, not a trickling stream, not a drop even of standing water was anywhere to be seen; the pumice-stone ashes had swallowed up all.

While I was still engaged in packing the photographic materials, a rumble was heard, and the poor boy made his appearance with a face of great alarm. He begged me to come and see! The western wall of the sailors' room, only a partition fortunately, had fallen down, and covered the floor,—where they had been sleeping the night before, and where he was to sleep that night,—with a mass of ruins. It was true the strength of the structure had been unduly tested by the yachtsmen, on their first arrival at Alta Vista; for they had tried slinging their hammocks, from pegs driven into mere dry walls. They could not, for a long while, understand anything being less tough, than the timbers and beams of naval archi- tecture; less still did they fancy, that a man could sleep comfortably, certainly not in ship-shape style, if his bed was spread on the ground, or arranged in any other way, than hanging from ring-bolts. Down however to the floor they were obliged to come eventually, to avoid the threatened catastrophe, always imminent, and only thus happily staved off, until some hour or so after their departure.

Besides this accident, however, one or two stones, and not small ones, had tumbled out from other parts of the walls, and for no perceptible reason, during the last few days; seeming to hint that our expeditious building, was not likely much to outlast the period, it had been so nicely calculated for.

The boy soon recovered from his alarm; and next, was only anxious to show, how much baggage he could carry up to the ice-cavern. Loading himself with twice the quantity I had set apart as his fair burden; he led the way untiringly up the angular passages of dark broken blocks in the Malpays, until we came in view of the "cross," that acts as the beacon or land-mark of this wondrous water supply, in a desolate wilderness of lava.

The cross so called, was only a thin single stick,— there had been once, we were told, a cross bit tied to it, making up a genuine *crux*,—yet still, mere small stick as it was, it loomed out an effective land-mark on those black lava masses. For in a region so purely of nothing but stones, this diminutive staff, was proof enough of human agency, and was not to be matched or mistaken for anything else found there, within miles and miles.

Down, therefore, in a hollow formed of loose stones, and then up upon a ridge of other loose stones, we went, making towards the stick. Suddenly amongst the most recent and fissured of all the blocks, where they seemed heaved up, and some of them tossed out and tumbling over,—there we came abruptly on the entrance to the ice-cavern,—a hole about three to four feet square, concealed until one was close upon it. (*See Photo-stereograph, No.* 13.)

An ugly place to look down into; for not as usual

with most caverns, was the entrance here on the
floor, but, in the roof; and that some twenty feet
high, above the rocky bed inside.   Hence it occurred,
that the first evening of our Alta Vista campaign,
two men having been sent up immediately after our
arrival, for a supply of water, and having reached the
locale about sunset—they looked around on the wilderness of lava, then down into the gloomy hole, and
became so much alarmed for their safety, as to defer
venturing into such savage depths, until the next
morning's sun should rise and gladden the scene.

Even then, the task was not very agreeable; for
one man had to descend, swinging by a knotted rope,
until, having reached the bottom, the casks were
lowered to him.   He then filled them with water, of
which there was quite a large pond full, but not vertically under the opening.   After the casks had again
been drawn up,—then came the tug of war in getting
the man out; for when he had at length got his
hands and arms upon the rough surface of the top of
the roof, there was no support, but the dangling rope,
to give a purchase to his feet.

On account of these *désagrémens*, the ice-cavern was
seldom entered before our day, but by those, to whom
it was necessary; viz., the hardy *neveros*, or peasants

whose trade is to convey ice and snow to the lower country. Ordinary tourists, we were informed, contented themselves with standing over the hole, looking down at the snow, and proving the fact of water in the gloomy space beyond, by throwing in stones; or, as there are few sizeable ones immediately about, one of their sticks; a large collection of which has accumulated there.

To remedy the inconveniences of descent, our yacht carpenter constructed and fixed, so famous a ladder, that a lady was handed down with ease, as well as the photographical machinery for wet collodion plates; and we left the structure behind, as a gift to the *neveros*.

On alighting at the bottom on a heap of stones, the first noticeable feature, was a ring fence of snow, three feet high and some seven or eight feet broad, extending nearly round our footing-place; while beyond, were large surfaces of water, stretching away into the further recesses of the cavern.

Amongst the multitudinous features of interest in any natural curiosity, no two persons may coincide as to which is the important one to be studied. Hence I could hardly agree with those who had suggested in England, that the wonder to be inquired into

here, was, how came there to be perpetual snow inside
the cavern, when there was none outside? The
rationale of this appeared at the place, by no means
difficult of solution; and thus. The *locale* of the ice-
cavern, abundantly covered with snow every winter,
is nearly within the perpetual snow line; so near, that
in its neighbourhood, and on the southern side of the
Peak too, we had seen patches of last winter's snow,
existing up to the middle of July, though exposed to
both radiation and temperature; the former, according
to our measures at Alta Vista, amounting at this
height to 130°, when the latter was only 47°. Within
the ice-cavern therefore, radiation being cut off by a
roof of rock, its contents have only to withstand the
temperature of shade; and thence can last so much
longer, than what lies outside, as to have something
unmelted to show through the whole of the year.

A similar result, from a like cause, was seen by Mr.
Airy this last summer in Casa Inglesa, on Mount
Etna. He had ascended all the bare slope of the
mountain, looking only for protection from heat, when
on entering this elevated cottage, he found its rooms
half filled with snow of the previous winter. It had
drifted in through an open door, and had been preserved by the thick roof from solar radiation.

A closer approach still, is to be seen in the Surts-hellir grotto of Iceland, as represented in the un-equalled lithographs of Paul Gaimard's large work on that country. One of his views represents the interior of a lava cavern, with a little hole in the ceiling, like a skylight; and under that, a hill of drifted, half melted, half consolidated snow; which is preserved through the summer, as though in an ice-house.

All this is so far precisely a representation of the cavern on the Peak, during early spring. An addi-tional feature, however, is needed, as the summer advances; for then, the central portion of the Tene-riffe mass gets melted out; partly, by the pencil of vertical solar rays streaming down at midday through the roof aperture; and partly by a modicum of vol-canic heat, which will presently be alluded to.

The inner sloping sides of the white mass, are stratified horizontally; or, in "elevation-crater" fashion, with veins of dust. These particles appear to act to some extent, as shields against effects of radiation; for between every pair of the bands, the snow is eaten more deeply. This was in so far, a phe-nomenon similar in principle, to the rocks found by Agassiz, Forbes, and others, riding on pinnacles of ice, and raised high over the melting, and continually lowering, surface of a glacier, during summer. In the cavern, how infinitesimally small the scale! Nevertheless, there was little more, whereon to allow this tiny patch of drifted snow, being called, as it was by a celebrated traveller, " an underground glacier." Our knowledge of these ice streams, their localities of origin, their mode of life and habitudes of motion, has so greatly advanced, since the above title was bestowed,—that there can hardly be occasion for occupying more time in formally objecting to such an application of the term. Let us rather hasten on to a greater wonder of the place.

What is that? Why to our minds the surpassing mystery was, how,—amidst the disintegrated material of the Malpays; the stones, stones, nothing but loose disjointed stones so often spoken of,—how, in such a locality, came there to exist a large and water-tight cavern? That is, in precisely the most unlikely part of a mountain, eminently unfit in its principles of constitution to be a retainer of water, and found in practical experience to be a mere sieve, and abominably dry on the surface; by what agency, in such a place, was formed this large and well-closed chamber with floor, walls, and ceiling?

YOUNG DRAGON TREES AND DATE PALM IN A CACTUS GARDEN NEAR OROTAVA.

p. 405. 413.

Printed by A. J. McKeowhy under the superintendence of James Glaisher, Esq. F.R.S.
and published by Lovell Reeve

Standing on the heap of stones, under that roof-hole whereby we had entered,—with the circular hill of snow around us, we looked upon a broad surface of underground waters, clear as crystal, that stretched away, far on every side. We were lost in wonder! The water itself could easily be the accumulation of melted drift; but how was it retained in a region of loose and open stones?

We did find on examination, that underneath the water, there was generally a depth of two feet of ice; but all the plugging was not done in this way. Moreover, what meant the smooth interior of walls and roof, rounded towards the top like a hollow dome smoothly plastered, and extending in three opposing directions, into long and gradually contracting lobes or conical channels? This structure must have been preliminary to the ice laid down on the floor.

"Oh! there is smoke coming out of the ground," exclaimed my wife. Sure enough there was a steady little column of some vaporous dusky matter, rising like smoke from a pipe; but quite certainly, and coming up fresh and fresh, in spiral vortices, to the height of two or three feet. It rose in the midst of a heap of stones, under the roof-hole, and exactly where the centre of the snow-mass had melted away.

Had some one thrown a cigar down there? that could not be, for our attendant boy had gone back to the station some time before this appearance manifested itself. Besides, he had not shown any smoking materials; we had none; and there was no other person on the Peak of Teneriffe that morning.

I laid down on the ground, trying to make out the particular place of exit for the vapour, but had some difficulty in identifying it. The smoke, if such it was, worked its way through the pores of the loose material, without sensibly displacing or changing it; and after a quarter of an hour, the appearance vanished.

We then went on examining the interior. In some crevices of the roof, was a beautiful and intricate grey fretwork of stone; not pumice, nor scoria; with nothing of the vesicular look of cinders, yet highly porous; and formed apparently by the dissolving action of water, by rain and melted snow, for ages finding their way into the cavern, through the same crevices, and washing out the more soluble portions. Often the honeycombed masses would be coated with a minute green lichen, that gave them much the colour of some of the magnesian materials of the "lunar rocks;" but the vegetable origin,

and superficiality of the covering, were soon dis-
cerned.   Though too, the crevices in the roof had
long since passed through all their annual stock of
snow-water, those at the further and deeper end of
the long western branch of the cavern,—that penetrated under a lofty ridge of lava, rolled down from
the Peak,—still kept on dropping water; showing
how the wave of summer heat, was slowly penetrating deeper and deeper into the mass of superincumbent stones, charged with snow of the previous
autumn.

From every part of the surface of the mountain,
downwards, there must be an immense quantity of
this infiltration of moisture always going on; for a
great depth of rain, frozen or thawed, falls every
winter on the Peak; and yet its sides are unscarred
by water-torrents.   On some of its steepest slopes
exist still, as at Alta Vista, large surfaces of loose
pumice-stone sand, formed myriads of years ago, and
apparently waiting to be washed away, by the first
surface-stream.   No such cataract, however, forms;
inward only, and still more deeply inward, seems to
soak all the rain-fall of this huge cone of loose
rubbish, this "eruption-crater."   The heaven-born
supply is lost to man, except the little that percolates

into the ice-cavern. That at least is secured; for whatever comes in there, goes no further.

The purity of this water is so admirable, that in photography I could use it freely, as distilled water; and the quantity is so great, that without improvising some sort of raft, one could not measure the size of the cavern. Its general figure, as already mentioned, is triangular; like a cast of a three-rayed star-fish. Seventy feet perhaps may be a fair guess at the length of the larger of the three lobes, up to where its continually declining and narrowing roof meets the level of the water; fifty, and forty, feet measuring the other two.

As we sat with the photographical tent on the central heap of stones, hour after hour, now trying a plate of the elevation crater-like snow-ridge; now the curving sides and branches of the cavern; and now the masses of rock, fallen from the middle of the dome-like roof,—a shower of rain swept over the mountain outside, but disturbed not the deep stillness underground. That was broken, by nothing but the quiet drop dropping, now singly, now harmoniously combining, of the water drops, that was always going on at the farthest end of the cavern. Often and long we gazed into those three dark, receding, conical hollows, opening upon, and directed on us, as we sat there in the midst.

Some grand and mysterious working of nature, must, we felt sure, be connected with their forms; and the story would be wonderful could it be successfully interpreted.

Caverns of one sort or another are very common on volcanic mountains, and very curious ones of a tunnelled character, are often found in lava; being caused by the hardening of the crust over the end of the stream, and then the sudden breaking out of the lower part, from hydrostatic pressure, by the interior fluid. Thus relieved from confinement, the fiery deluge makes another stage down the mountain side, until once more casehardened by radiation of heat from its surface; again to break out, when the internal column of fluid lava, continually being added to from the crater, rises up to, and beyond, the resisting powers of the crust. The Surtshellir gallery in Iceland, already mentioned, is one of these; and another of enormous size is well described by Wilkes in his ascent of Mouna Loa, in the Pacific.

With their tubular character, and their horrible mouths yawning downwards, the appearance of this

class of caverns is unmistakeable; and the ice-cave on
Teneriffe is none of them. A lava stream has certainly issued from its site, but not out of any of its
trinary tubes.   The two that are not absolutely
pointing to the mountain's top, are neither of
them turned precisely downward, or the direction in
which the lava actually flowed.   In place of this, one
tube is at an angle of about 50° on one side, and
the other at an angle of 70° on the other; and both
are at a great depth below the surface.   The third
tube lies towards the Peak, but as its axis is inclined
downwards, the lava from above could hardly have
flowed through it.

From the general site of the ice-cavern, however, a
fresh outburst of lava seems to have taken place,
notwithstanding its situation in the midst of the
Malpays, or far below the rim of the great Rambleta
crater.   This is in itself an important point to establish; and proofs seem abundant; for we find a
prevailing convexity over the place, as clearly marked
on the upper side towards the Peak, as on either
flank.   Then on the lower side, a distinct ridge descends from it, and the further we follow this, the
more rolled do its materials become; until at last
what had been sharp angular fragments of large size,

with unfilled interstices, become heaps of rounded stones, with *debris* of pebbles and dust amongst them. (*Compare Photo-stereographs, Nos.* 10 *and* 13.) This last stream in its day, must in fact, like the Vesuvian specimen described by Sir Henry de la Bêche, have degenerated in its downward flow, into a mere ridge of clinkery cinders, rattling and tumbling over each other, as the general mass continued to advance.

In the immediate neighbourhood of the ice-cavern, the lava masses are particularly clean; and for a considerable space, have their upper surface covered with a sort of felt-like substance, arranged in parallel ridges; at a little distance, looking strangely like the plaiting of the interior coats of a stomach. (*See the front blocks of Photo-stereograph, No.* 13.) On closer inspection, the matter is found to be hard, though porous; and is a sort of frothy scum, due to a very bad attempt of this stony lava to form pumice.

Over the actual area of the cavern, this tripy surface, if we may so call it, has been much broken up; but apparently only by one great throe, since which, it for ever became still, and has lain precisely as we find it now. This appears from the circumstance, that wherever a portion with the original rough ridges, is wanting, there is a square angled hole in the dense rock; and the missing fragment is mostly found lying close by, and overturned. One such mass, of particularly large dimensions, is to be seen towards the south, completely upside down. With its clean and broad fractured side, once the lowest, now upper- most, it serves a purpose that its former ridgy and scoriaceous surface would have been most inappro- priate to; for it makes a famous flat and level table, whereon the *neveros* from time immemorial have packed, and made up their parcels of cavern snow, before venturing to carry them under a vertical sun, through the thirty miles that separate the mountain from the capital town of the island, (Santa Cruz).

Tracing round and round the exterior of the cave, we found portions of the rock, with the curious tripy substance over the upper surface, thrown outwards, and to every quarter; even towards the higher or Peak side. In this direction too, a stream of lava descending from above, has evidently felt the resistance of the ice-cavern crater of emission; and instead of descending at its former angle of about 25°, has accu- mulated into quite a steep head over the place. Were further testimony needed, I might go on to describe the resemblance between the commencing rocks and fissures of this ice-cavern stream, and those that break out so abundantly from every part of the rim of the Rambleta crater above. But enough must have been already given, to prove, that there has been here, over the site of the cave of ice, a fresh escape of lava from the interior of the mountain.

We ourselves at least were satisfied; and on descending into the pit once more, we instantly recognised in the smooth rounded surfaces of its interior, the action of highly elastic gases on a plastic material; and in the three conical mouths, so many volcanic vents; trumpets as it were, blaring up from three different quarters under the earth; and in the place where they met, heaving up the lava into a dome-like elevation, by their united forces.

The long stream of stones below, and in connexion with this spot, shows that the first manifestation at the surface, when this special volcanic disturbance began, was the appearance of hot glowing lava. That was the alpha,—the formation of a cavern, and then the escape of its elastic vapours,—were the omega of the eruption.

Being one of the latest outpourings of the Peak, the early Guanches may have witnessed from their goatfolds, the origin and progress of this interesting labour of the mountain, and not unfraught with danger to themselves. The scene must have been striking. First a bright spot of fire appears in the Malpays, then a lengthening, red, and smoking line, increasing in breadth and depth as it flows. On it comes nearer and nearer, driven by some unseen power within the bowels of the mountain. Crashing rocks in its path, tell the weight of the fiery flood; slowly, though surely, advancing over every obstacle. From time to time a fresh wave of fire shoots upward at the fountain-head; and the whole stream feels the impulse, though the lower end is now almost hid under masses of blackened scoriæ; dismal blocks of coal-black stone continually hurtling, rolling, and falling over each other, with every onward motion of the fluid lava beneath. Occasionally a steeper slope intervenes, and then the full nature of the impending destruction is manifested, to the terror-stricken natives, in the cascade of alternate fire and dross that comes thundering down into the valley upon them—the stones of darkness, and the shadow of death.

At length nearly the whole column of fluid in the volcanic duct has been extruded. The tense and elastic vapours, that in their struggles for freedom, have been producing all this mischief—feel their

success and are quick to follow it up. Rejoicing in
the diminution of hydrostatic pressure, they rush
upwards; the three vents meet together; one collected heave to gain the light of day, and up goes the
great lava bubble, the expanding space occupied by
the compressed gases. A great heave; but the cold
of the atmosphere has done its work outside—the
mass is hide-bound with frozen stone; the thrust,
however, becomes irresistible; before its violence the
strong coating is rent and shattered into a thousand
pieces; some of them are tossed out afar, and the
whole mass lifts up for a moment, in disrupted
portions.

That instant is the triumph and destruction of the
power of these confined gases; the cavern already
formed, has its upper surface partially blown off: but
at the openings which they have thus made on every
side, the vapours flash out, like exploding steam of
highest pressure; they are dispersed into thin air,
and the heavy sections of the lava roof, unsupported
from below, fall downward again, wedging into and
against each other, so as nearly to reform their
previous figure.

So must have escaped these unruly spirits from the
interior of the mountain; and then there was quiet.

Had their elastic force been much greater, the whole roof of the cavern would have been carried completely away, and men would have acknowledged the aperture as a crater of eruption. Had the lateral extent of hollow been greater, or the thickness of its cover less, and the elastic power no more—then would have resulted after the explosion, a complete crumbling, and falling down to the bottom, of all the central region of roof. In such case there would have been left, a great heap of rubbish in the middle of the floor; accompanied with a concealment of the long trumpet hollows. Thus would have been formed indeed, a sort of central pit, bounded internally by the precipitous fractured edges of the roof, and externally by the gently sloping sides of the lava mound. Had all this taken place, as it did very nearly and even has done partially, then would men have been equally ready to claim the spot as a crater of elevation; that much lauded and much abused, stoutly asserted and more denied, magnificent generalization, by the learned Von Buch.

To those, however, with whom principles weigh— the degree to, or the scale on, which effects have been carried out, is not so important as the facilities which a case may present, for testing the law which has been exerted; and that depends rather on the opportunity offered for making accurate observations. Now precisely in this, is the important use which the ice-cavern may subserve; for we may therein descend under the surface of the ground, and minutely examine nature's mode of volcanic operation, in secret recesses usually hid from man! There, unblocked with rubbish, the passages by which the condensed vapours effected their junction, and their combined explosion, are as perfect as on that distant day, when their sides were soft with heat, and no human being could approach without destruction. Protected by the vaulted roof—from falling stones, from dust and ashes, as well as from the rending and destructive power of alternate frost and sunshine,—the walls of this volcanic house, are preserved underground in all the freshness of the buried palaces of Nineveh. The only change that has occurred with the lapse of ages, is the entrance, every winter, of a modicum of drifted snow, and the appearances occasioned by its melting, filling up, and standing in the three descending branches of this remarkable triangular cavern.

On mentioning subsequently to a native of Teneriffe, this theory of the volcanic origin of the present ice-cavern, his countenance grew very serious; and on speaking further of the little gyrating column of vapour, seen rising from the heap of stones in the centre of the floor—in a manner and position conformable with the growth of a crater of eruption, —his face grew positively long, and he looked about him, and over his shoulder, rather uneasily. We were still on the flanks of the Peák, amongst scoriæ and ashes. By and by, however, he plucked up spirit, and began to talk of lots of caverns about the mountain; long well-known and not feared at all. There was one near Icod, he averred, on the northern coast; it was quite safe; and communicated with the ice-cavern. What, all the way from the coast eight miles off and 11,000 feet below? Yes, the whole way.

After we had returned to Orotava, I could not but recur to this story; and I fear must have expressed myself somewhat depreciatingly of its importance; for all the company present, immediately spoke out in its defence. Several worthy and reverend hidalgos, they said, had lately met together to prove the truth of the general belief, had driven a dog into the entrance of the cavern at Icod; and a few days after, some *neveros* found him lying, greatly fatigued and emaciated, on the floor of the ice-cavern on the Peak.

I could then of course only wonder, at how the dog in his subterranean climb, could have entered any of the three water-logged descending branches of our ice-cavern, without letting all the water out upon himself; and being carried by the rush, right down to Icod again, and possibly drowned in the sea.

CHAPTER IX.

LAST OF THE MOUNTAIN.

IN descending from the ice-cavern, along the very
ridge of lava that had issued from its site, the
gradual change in aspect of its material was interesting; not only from the rolled and worn look of
its earlier and now lower portions, but from the
alteration of colour and quality. At the cavern
itself, the substance was of a bluish grey in the mass,
amazingly tough and of basaltic consistence; while
lower down the stream, it became black, then
brownish-black, somewhat cindery and brittle; this
portion having been indeed the scum that the volcanic vent had first poured out.

Amongst other fragments met with, were occasional specimens having one or more sides as smooth,
as if they had been polished. This was no result
of simple fracture,—for of all the blocks which we
broke by hammer, there was not one that did not
show a rough, granular, though sometimes glossy

YOUNG DRAGON TREES (DRACÆNA DRACO) NEAR OROTAVA.
p.414.

Printed by A. J. Melhuish under the superintendence of James Glaisher, Esq. F.R.S.
and, published by Lovell Reeve.

surface. We were puzzled by those smooth natural faces, for they neither arose from the material being vitreous, nor from any result of rubbing and grinding; but of this we were quite certain, that they served admirably to display the laminated nature of the lava.

Perceiving presently a heavy lump, rather bigger than a man's head, excellently illustrative of the contortions of said laminæ, we picked it up as a prize specimen of the first class; but had no sooner done so, than the Canarian youth attending us, though loaded already with a box of chemicals and a photographic tent, prayed for his right of place to carry it.

Again a little further on, we selected another characteristic mass, something so large as to be proper for a geological, rather than a mineralogic, museum; and again he almost insisted on his duty as bearer; nor was he content, until the weight of stones he bore, in addition to the rest of his burden, was quite a caution to behold.

What a pity that Humboldt in his visit to the Peak, had not the services of such an attendant. In the "Personal Narrative," p. 72, he complains sadly at this very portion of the journey. "Un-

fortunately," says he, "the listlessness of our guides
contributed to increase the difficulty of this ascent.
Unlike the guides of the Valley of Chamouni, or the
nimble-footed Guanches, who could, it is asserted,
seize the rabbit or wild goat in its course, our
Canarian guides were models of the phlegmatic.
They had wished to persuade us on the previous
evening not to go beyond the station of the rocks.
Every ten minutes they sat down to rest themselves,
and when unobserved they threw away the specimens
of obsidian and pumice-stone, which we had carefully
collected."

The Baron indeed, must either have been remarkably unfortunate with his men; or the Islenos have
improved extraordinarily in the last sixty years;
for all of our attendants, Manuel "the Marquis,"
Manuel the guide, and Pédro, and José, and Márcos,
and many others, were "models" of mountain-
climbers; never making any difficulty about distance
or weight, up-hill or down-hill.   They even ascended with remarkable ease, notwithstanding their
loads, the terminal sugar-loaf of the mountain,
which Humboldt himself did not get up very quickly,
and which he describes as the most difficult mountain
he ever essayed, except Jorullo in Mexico.

Let us hope that the explanation of the alleged incompetence of Humboldt's guides, lies in the subsequent moral and physical advancement of the Spanish inhabitants; for in the present day, when we are hearing so much of the gradual degeneration of Saxon and Scandinavian, it would be so pleasant to think, that a nation, to whom Europe owed much in early times of trouble, is again going ahead in those qualities that form an essential foundation for the effective greatness of a people.

Another rock specimen still, I wanted, but let the brave Teneriffe lad get a long way ahead, before venturing to pick it up. It was a large lump of white and yellow felspar, the material of the terminal crater; and it was lying on a red pumice platform, immediately below the black lava streams. How came it there, so far horizontally from its place of origin? There would be little difficulty in accounting, could full credence be given to the tales collected by a noble and clever author of a recent visit to Iceland, setting forth that in the terrible eruption of Skaptar Joküll, in 1783, great rocks were thrown to distances of fifteen and twenty miles off! Until ocular testimony of such lateral projections are brought forward, we shall do well to confide in the

unvarying accounts from Vesuvius, Etna, and Chajorra itself, that a chief feature of the stone-shooting
of a mountain, is extreme accuracy in vertical flight;
so that most of the projectiles fall back into the
crater again, or close outside.

Mr. Stephenson, who watched the stone-throwing
of Vesuvius with much care, concluded, he told me
lately, from their regularity and other features, that
the effects were due to water. A portion of this
introduced into the duct of melted lava low down,
he thought must be heated to a degree equal to both
the latent and sensible heats of steam, of great elas-
ticity, though not equal to the pressure of the column
of melted stone above it. Up however, by reason of
its small specific gravity, rises the globe of red-hot
water, through the heavier liquid; and when at length
it comes so near the surface, that the tension of its
steam is greater than that of the decreasing column
of lava, an explosion takes place; and surface-masses
of scoria are shot up from the well of the crater, as
though it had been a cannon sunk in the ground,
and with its mouth pointing accurately to the zenith.
Closer still to Skaptar Joküll, is the Great Geysir,
whose columns of water, though rising from 60 to
120 feet high, never diverge, but go straight up, and
fall nearly straight down again.

With these data then, it would hardly be safe to conclude anything more startling, than that the pieces of terminal crater which we found near Alta Vista, must have owed their horizontal displacement, to rolling down the side of the mountain; before it was endued with its present roughness, consequent on the emission of the last class of black lava streams. On Alta Vista, at all events, we were upon an uninterrupted line of sloping descent from the Peak; and the thing was possible. But on the top of Guajara, with a depression of upwards of 1000 feet, between us and the Peak, or for the fragments to roll up, we felt ourselves justified in denying, that certain things shown to us there, were "bombs" from the Piton.

Returned to our lonely station, we began to make ourselves comfortable before night set in, the last we were to spend at Alta Vista; and a night so uniformly cloudy, that there was nothing either scientific or artistic to look at. With no troubling conscience therefore we could turn our attention to the culinary department, not a little curious in some of its details, at a height of 10,700 feet; and requiring some personal attention, as our small expedition was unable

from first to last, to boast of either a regular cook,
or any one set apart for pure serving purposes.

Raised above the limit of vegetation, we could no
longer at Alta Vista indulge in the blazing fires of
Guajara, and those refinements of fuel, from its
retamas and *codesos*, which there lay spread about
on all sides so invitingly. Fuel at Alta Vista was
indeed too expensive a luxury to be indulged in
much; and though a mule-load of wood had been
brought up one day from the Estancia, it had got so
thoroughly wetted on the night of the storm; that
our men, taking it by turns to blow, had very hard
work in getting their coffee-pot boiled.

The Spanish lad therefore, all by himself, would
have had very weary work of it in procuring his
supper, had not I suddenly prevented him with a jug
of hot tea, quickly heated up over a spirit-lamp of
the first magnitude.

Coming out from England prepared to measure
the height of mountains by the boiling-point, we had
been provided, for the heating part of this operation,
with one of those inflamed vapour-jets, known as
Russian furnaces. Greatly it astonished the whole
camp at Guajara. Filled with proof spirit, it raged
and roared like a mad little demon, dashed its column

of hissing fire with such violence against the bottom of the saucepan, as to recoil on every side in the form of an umbrella of flame, and effectually deterred every one and any one from attempting to read off the exact height of the thermometer immediately above it. After this exhibition, the thing was degraded to kitchen purposes, and only employed there, when hot water was wanted in a hurry and at any cost. Generally a copper lamp, holding about a quarter of a pint of spirit, and furnished with a wick an inch in diameter, was found to be the most practical plan for boiling, either .the thermometer, or eggs, or any of the smaller sort of deer.

But when there was genuine hard work to be done, as stewing beef—tough muscles of beef which had been half-cooked in Orotava, and since their arrival on the mountain, had hardened into something like ebony, from the aridity of air above the clouds,—then the business was done by one of Price's candle lamps. Splendidly too, for as the pot was thereby kept simmering, without cessation for eight hours, what meat so hard-hearted as to resist the insinuating appeal.

At the height of 10,700 feet then, we had splendid stews—meat rendered soft, fat floating about on the top, and all that sort of thing,—so of course we could and did cook eggs, either soft or hard as we chose, with a boiling temperature at no more than 193°, accurately 193°·32 in a mean state of barometer. When, therefore, a clever author at Erzeroum, only 7500 feet high, describes that eggs took there, he did not know how long, to have merely their whites fixed; and that boiling water did not scald him,—he clearly did not manage the fire as effectively, as he assuredly has done, his flowing pen. When further, he describes that his tea at the same place was always vapid and bad; that there was " bubble, bubble, boil and only trouble," on account of the low temperature of boiling water, by reason of the great altitude of his *locale*,—plainly the fault must have been his own, for he tilts both against theory in general, and our Alta Vista practice in particular.

The flavour of tea, as of all vegetable infusions, chemists tell us is finer and purer, less loaded by coarse, acrid principles of stalks, seeds, &c., when cold, than hot, drawn. But unless we use hot and even boiling water, the infused leaves, charged with air, float about unpleasantly in the fluid. Now on the mountain, being able to secure the expulsion of air by complete boiling, at a temperature nearly 20° lower than 212°,

we had by so much a more delicate and refined preparation from the leaves. Such is the theory; and
our practice fully confirmed it, for my wife, who
always makes good tea, excelled herself at Alta
Vista.

There could be no dawdling, however, over a cup,
after it was once poured out; for what with the low
temperature of that boiling water to start from, and
then its rapid evaporation under a barometric pressure of little more than twenty inches,—the contents
would become cold, before a prosy speech was finished.

This was about the utmost practical inconvenience
that befel us from rarified air; for as to the alleged
sickness, feverish pulse, bleeding at eyes and ears, &c.
&c., we felt nothing of the sort. Yet we were told of
gentlemen ascending the mountain, and being taken
so desperately ill, on arriving at Estancia de los
Ingleses, 1000 feet below our station, that they were
obliged to give up the object of their ambition.

How are such anomalies to be explained; and are
they the legitimate effects of a rare atmosphere?

If a windlass, or a treadmill were erected in London,
and a gentleman in easy circumstances set some fine
morning to perform at one of those ingenious
machines, an amount of work, equal to the mechanical

task of raising his own weight up through the height of 10,000 feet perpendicular in seven hours,—I believe that, though breathing air, of a density of thirty mercurial inches, he would be distressed as much as the traveller who, by ascending a mountain, performs the same "duty," for the first time in his life.

On September 19th, the day fixed for our leaving the mountain, the sun rose dimly behind watery cirrostrati, that covered the whole of the sky; just in fact as they had done every day, since the storm that ushered in autumn. on the heights. A few lower clouds appeared also at the 4000 foot level; but well appointed cumuloni no longer, they wandered about uncertainly in ragged fragments, rolled to and fro by every motion of the air. The N.E. Trade, the characteristic of a Canarian summer, was gone, and the S.W. wind, though blowing on the mountain-top, had not yet descended to take possession for the winter of those lower zones; where a meteorologic interregnum was in consequence evidently prevailing.

The air, however, was exquisitely transparent, and as we gazed downwards on our visible half of the great elevation-crater, we could hardly disabuse ourselves of a false perspective, which made the distant, appear to rise above the near part, of the plain.

Beautiful little craters dotted the surface far and wide,—not like the too ordinary ideal of a terrestrial volcano, a high cinder cone with a little hole at the top,—for they were pits or basins, often so broad in comparison to their height, as to look like cups or punchbowls. These were casual decorations; the more essential markings of the surface, were lava streams that had descended from the Peak. Towards the east and south, they had rolled in great waves, and then stopped, there was no egress there; but in the N.E. quarter, they were seen in converging flow, making for the head of the valley of Taoro. Through that exit, we knew that they had poured headlong and tumultuously; for had we not been there; and in their broken remnants found the vesicles so common to lava, pulled out into long thin tubes by the downward flow of the pasty mass. Yet though we knew all this, such was the clearness of vision for distant things, that the " gate of Taoro " on the crater's verge, looked to us certainly higher than that central portion of the floor, which lay close under our feet.

" Those converging ridges prove that lavas most

assuredly poured out there once, though they could
by no means do so now, from the subsequent bending
in of the centre of the floor." Thus we were induced
to argue on Alta Vista, and little thought that the
barometric observations on our journeys would show,
that in every direction in which we traversed the
crater plain, there was a very sensible descent from
centre to circumference. Not only, therefore, did the
ancient lavas flow out by the "gate of Taoro," but
the fall of the ground is still such, that new lavas
would follow the old, were large streams once more
projected in the same direction.

Such projections and emissions, through ages, may
have formed the chief central mass on the plain; culminating in the midst of which, is the Peak with its
many mouths. Out of these, agglomerating materials have certainly been vomited, rolling down the
flanks, and continually increasing the size of the
heap. The Peak is therefore, in so far, undoubtedly
a "crater of eruption." Every one allows this, and
consequently no cavil will be made as to the title of
our Part III, or "On the Crater of Eruption."

But there may be a contest on the propriety of the
"crater of elevation," which we have inscribed as a
heading to our Second Part. Not but that every

geologist will know perfectly well what it means, for of course he will understand that it refers to Guajara, to the Somma of Teneriffe; to the great crater, that surrounds the Peak of eruption, and is distinguished by the gentle slope and smooth surface of its external walls, as well as by the stratified nature and lateral expansions of its component lavas.

Yet notwithstanding, how many geologists are there, especially in England, who will loudly assert, that Guajara cannot be an elevation crater, because say they, there is no such thing as a crater of elevation in nature. In London this party has nearly carried everything before them; the fight rages nevertheless in other quarters, and with different success, particularly among Continental schools.

Being ourselves visited the other day by a great European traveller, we had an opportunity of hearing opinions from abroad on this disputed question; for the gentleman,—a worshipper of Von Buch, the inventor, and an intimate friend of Elie de Beaumont, the chosen knight, of the "*soulèvement*" theory,— held forth at great length on that side; and almost danced with joy, at some things that we were able to communicate.

What, said he, you have found a portion of the

great crater wall northward, as well as southward, of
the Peak? Yes; on the northern, or Pompeian side,—
where Von Buch, in his beautiful map, has only a
uniform slope,—there we found a considerable length
of internal cliff, precipitous and basaltic; with a
curving valley of pumice below, the Canada del Icod,
and forming as satisfactory a portion of the ancient
ring, as Guajara itself.   Behold too, we added, these
specimens of several strata of the northern wall, and
these of its southern counterpart; you will see that
they are nearly identical; and in these two photographs of ravines on either side of the great crater,
but eight miles apart, you may note the appearance
of the same remarkable bands of white tufa, at equal
but opposing angles.   Oh! this is really glorious, said
our friend; for this is precisely what will enable me
to prove, that the crater is one of elevation. So having
taken notes, he hastened off; and with his supposed
proof under one arm, and his umbrella under the
other, went running along the pavement, in such
anxiety to attack, or as the French have it, " to
tear down the face" of, some one of opposite
opinions.

Poor gentleman! those are facts that you have in
your hand; proofs of a portion of the northern circuit

still standing; but that will not by itself incontestibly
prove the elevation hypothesis.   Nor if you could get
testimony of seven portions of crater wall to complete
the whole ring, of which Von Buch chronicled but
half,—would you have any more title to use the term,
as long as other men can say with Mr. Jukes,—who
has visited the spot,—" the materials of these walls
are not elevated, but erupted."

"We know the vast size of the circle," they say,
the friends of Jukes, " we know it all, we have walked
over it, we have ascended its walls, we have chipped
its rocks, and can positively assert, that if they are
denser, more stratified and flattened out than those of
the Peak, it is because they were erupted under sea,
not under air.   Call it a *submarine* crater if you please,
and the Peak a *sub-aerial* one, but use not the term
elevation, for such force had neither act nor part in
the matter."

When this is the language, and these are the
conscientious convictions of good geologists and
mineralogists, after careful examination of the ground,
it is clear that something else, besides merely finding
additional portions of wall to fill gaps in the circle,
if that circle was erupted,—must be brought up to
prove the fact of elevation in Von Buch's sense, or

a swelling up of strata from violent action by compressed gases below.

Though our friend would not see it, we thought
this Jukesian view a heavy blow, and sore discouragement to his views; and we could not forget that an
ingenious teacher in the science, says, that the only
known method of volcanoes forming or growing, the
only way in which they are *seen* to increase, is by
eruption; and that it is contrary to the principles
of true philosophy, to employ an unknown force,
" elevation," to account for effects that are capable of
being produced by the known force, " eruption."

Much truth is contained in this passage; but then
what a state the author is left in, when we can point
to the ice-cavern of Teneriffe,—as in principle and in
extent, as far as it goes,—a pure elevation crater, and
of very modern date.

We have there, undoubtedly, the surface of the
ground heaved up by the force of dense vapours
below, and then a partial falling in of the topmost
part of the elevated stratum. Had the same forces
had their *locale* in the bed of a deep sea, instead of on
the side of a mountain,—the first exuding lava,
instead of running down the slope, would have spread
far and wide over the duct, as a level stratum.   Then

when the gases came up at last, there would have been a magnificent bubble or hollow dome of stone. This, on ultimately cracking with the tension, must have allowed its aerial contents to escape; when immediately, the superincumbent fluid pressure must have broken in the top of the cavernous hollow, completely and with no mistake.

In such case there would have formed, without doubt, a "crater of elevation," of very perfect character; but it could not be more exact as to principle, and certainly not so easy or satisfactory to examine, as the present ice-cavern, found high up on the sub-aerial Peak.

That ice-cavern then settles, that "elevation" *is* a force, and a method, in Nature. Once exerted, a small amount of erupted materials thrown out afterwards upon the site, and lying on the top, cannot annihilate the older fact. Consequently, if there are any direct evidences of such preliminary elevation in the great crater of Teneriffe,—we may freely allow to Mr. Jukes, that the present surface walls are of erupted materials; and notwithstanding that, we shall still be perfectly justified in calling the general structure, by the name of its first and grandest action, a "crater of elevation."

To this end then, only take up a map of the island;
look at its triangular figure; turn it round 180°, so
as to make the longer end point westward—and it is
a plan of our ice-cavern, with its three branches in
their relative proportions and proper directions. Do
not attach any immediate importance to this, but
study the "Lunar Rocks," and more especially, three
sets of cliffs at the root of the wall in the south-east
corner of the crater. They are given in some photographs taken on Guajara, and are portions of the
upper part of the crater wall, that have fallen; and
how? They have not fallen prostrate, measuring
their length on the ground, nor have they tumbled
disintegrated in a mass of rubbish; but each one, as
a large sectional column from the face of the cliff, has
loosened its hold of the rest, and dropped vertically
right down into the earth, on the very spot where
once it stood.

This is indeed a most remarkable feature, and undoubted. In one case, where there is only just visible
above the surface of the ground, a few feet of what was
once the summit of the crater wall, there is no *débris*
about, representing the ruin of the lower part of the
column; and there is no stream of water that could
have washed such softer matter away; there is indeed
rather a hollow at the spot. No other conclusion can

DRAGON TREE WALK AT A PALAZZO NEAR OROTAVA.

p.414.

Printed by A. J. Melhuish under the superintendence of James Glaisher, Esq. F.R.S.
and published by Lovell Reeve.

therefore be come to, than that the lower end of the mass has gone down into an empty space below!

In our "road of the Guanche Kings," is a proof that Guajara is split across, parallel with its face, from one lateral ravine to the other. If that part falls, as falls an ordinary cliff, immense will be the mound of rubbish; but if there is some hollow underground ready yawning to receive it, the mass will simply drop down bodily into the pit, and leave merely the tops of its precipices, if tall enough, above the level of the crater floor.

Such has been the fate of many portions of the wall. They are lowered from their once high position, but are not altogether lost to the light of day, and the eye of man. To the eastward, however, a greater calamity has occurred. There,—over what may very probably be, from the external figure of the island, the longer and larger subterranean tunnel, —has been an extensive bending down of the strata, even outside and beyond the crater circle; and where this bending surface does at last meet the crater, there is a large portion broken out, and gone.

The appearance from the Peak was most unmistakeable; there was an evident conchoidal fracture. At the place itself, we found a great bay or recess in the crater wall, with a level floor of pumice-stone sand. Where has this missing part of the wall gone? Where, but into the great hollow below; that hollow which measures the rise of the surface rock, and has made this eight mile crater of Teneriffe, a genuine crater of elevation.

The celebrated author of the " Principles of Geology," does indeed hint (seventh edition, p. 419), that Teneriffe cannot be a crater of elevation, because, " its opposite walls do not correspond in such a manner as to imply by their present outlines that they were formerly united," and because, " the precipices on opposite sides of the crateriform hollow would not fit if brought together, there being no projecting masses in one wall, to enter into indentations in the other."

But such want of fit in the opposite sides, is precisely a proof of a former elevation by gaseous force; because in such a case, large breakings in, of the centre of a raised crust, when the scale is very extensive, must inevitably occur. In Teneriffe, accordingly, not only was there a great hole, broken in, at the instant of the escape of the gases, but the sides of the pit so formed, have gone on cracking, breaking down, and falling in, long after the grand catastrophe; even after the crater had become sub-aerial.

A fearful picture might indeed be drawn of the

probable state of the interior of Teneriffe; the volcanic gulfs which exist under the fairest regions of the island; the "elevated" walls of the great crater, almost cracking with their own weight, as they hang suspended over an awful chasm; in midst of which,—amongst the tumbled ruins, of fifty square miles of country which fell in together to form the pit of the vast crater,—there is restlessly seething a tide of molten lava, rising from yet lower depths. Sometimes it rushes upwards with supernatural fury to the top of the cone of erupted materials, which it has thrown out in the course of ages, the Peak, and its craters Rambleta, Chajorra, and Montaña Blanco; then ebbing again, it retires for an unknown interval, into the central fiery caverns of the earth.

With our measurements, photographs, and drawings as data, we had even begun to trace the strata downward to some of the cavernous hollows,—but were appalled at the presentiment of evil; of a catastrophe, which all the power of the calculus would be utterly unable to predict.

Yet we have a hope, a great one, the secular progress of the physical world; obediently to which the fires of Teneriffe are dying out and the whole mass of the earth, inevitably cooling down.

PART IV.

—

LOWLANDS OF TENERIFFE.

CHAPTER I.

SEASONS AND PLANTS.

O N September 20th, in a luxurious temperature of
78°, at a table decorated with heaps of glorious
purple figs, richly ripe mulberries, bowls of fragrant
honey, cups of chocolate whipped up into froth displaying all the colours of the rainbow, &c. &c., we
took breakfast in Orotava, our minds in a state of
perfect ease that we had long been strangers to. And
why? every one was safe down from the mountain,
and all our goods and chattels as well. The last
mule loads had arrived without accident the previous
night, and were now under lock and key close to us.

The sun came out with almost oppressive warmth
amongst broken clouds; the air was steaming and
hot, feeling like a vapour bath; papau and banana
trees waved their fine fronds before the window, amid
passion-flowers climbing everywhere. The barometer
might be standing at thirty inches, but we did not
feel so capable or so inclined for exertion as in the

thin, dry and cold air of Alta Vista. From early autumn, we had suddenly returned to the height of sultry summer. There was more in this, than at first met the eye; more than mere difference of elevation would account for; and various changes had occurred since we were last here.

Thus, we found the N.E. cloud broken, the N.E. wind gone, exactly in fact as we had concluded when on the Peak; and in place of that closely clouded region that had dimmed all the distant sea in July, and August, we now saw blue sky and floating cumuli; while below them, the edge of ocean appeared dark purple, and distinct as a wall on the horizon.

We went down to the beach to photograph the surf, so characteristic a feature in July; but it rolled in no longer with its thundering roar; it had nearly departed, with the extinction of its parent Trade-wind.

Having been most liberally furnished, when passing through London, with meteorological instruments by the scientific Admiral Fitzroy,—we had been thereby enabled to lend apparatus to several spirited observers in Santa Cruz and Orotava; and now found, on examining the registers which they had kept up with most commendable zeal,—that, at the sea level up to

this day in September, the mean temperature had not
yet reached its maximum. Comparing this with our
mountain journals, an interesting result was arrived
at. On Guajara, as well as at Alta Vista, the maximum of summer temperature had been reached in the
beginning of July, since which time there had been
at both places, but most so at the higher, a well
marked decrease of heat. Autumn had begun there in
fact, in cold as well as in cloud. The simple conclusion was therefore, that the seasons are earlier, more
conformable to the declination of the sun, as we ascend in the atmosphere.

This view once obtained, we no longer were surprised at the progressive differences in temperature
between Alta Vista and Orotava, through August and
September, from 18° to 38°; for with the annual heat
curve of one station going up, and the other down, at
that time,—what other result could take place.

The fact was undoubted, and we instantly perceived the advantage of ascending high mountains
early in the season; for then, from spring below,
reaching summer above,—we may enjoy life in upper
regions of the atmosphere; without incurring any
very great change of temperature, dependent on the
degree of cold properly due to height. Poor Dr.

Wilde and his sick friend, in 1837, appear to have ascended the Peak in autumn, the autumn of the sea coast; and so, reaching in a few hours the winter of 10,000 feet in altitude, they experienced the horrible cold he describes, one would think almost as bad as that of the Polar regions.

Two or three cloudy days occurred after our descent; and then, the sky shone out transcendently blue; so I hastened up to the hotel roof, with photographic camera, to try a portrait of the Peak. From this point, at the very coast, we could see all the way up to the culminating summit of the mountain. Twelve thousand feet difference of elevation, and eight miles of distance appeared something trifling in this clear air; while the cliffs of the eastern flanks of Tigayga, strongly illuminated by the rising sun, showed a brilliant line of coloured faces of rock, marked here and there with as strong shadows. Wherever the ground was of gentle descent, tints of vegetation were perceptible; grey *retamas*, at the top of the ridge; vivid green heaths, as a stripe across its middle. We could actually distinguish the place, 5000 feet in elevation, where our men, standing on the edge of the cliff, in our descent of the previous month, had shouted with joy at the sight of Orotava, and the fertile lands below. The place was clear enough, just above the edge of the zone of heaths; but appearing so distinctly and so close, who could imagine it really as high, and far off, as a map made it out to be.

The senses of man might be mistaken, but not the photographic camera. Its first plate gave a strongly pronounced picture, the town of Orotava glittered in the foreground, and the Peak, cut sharply against the sky, in the distance; but what had become of that buttressing ridge of Tigayga, to the eye so strongly illuminated on one side, throwing so black a shadow on the other. Not a ghost of it, or its brethren, appeared on the collodion film. We tried another and another, ringing all the changes of long and short exposure, positive and negative developers, yet all to no avail; the detail of the escarpment would not come out. There was only the sky line, and a flat tint within that; as if the sun were behind, not in front, of the mountain.

Every photographer knows full well the difficulty of reproducing distances; and no one would expect, in England, to get the detail on the surface of a mountain, anything like four miles off. But we had just descended from Alta Vista, and had there, day after

D D

day, got pictures of the crater-wall at that number of miles from us, with every ridge, every stratum, and almost every *retama* bush marking itself in oxide of silver. At the time, we thought only of a southern climate; but now, in the same climate, at the same distance, with the same altitude of sun, we found that it must have been the elevation of 10,700 feet for the camera, and 9000 for the mountain operated on, which gave all the minutiæ of view to the collodion plate.

Saussure had, long before the days of photography, determined that the chemical power of the sun, increases directly with one's ascent in the atmosphere; consequently we had felt little surprise, that our mountain pictures were taken, generally, with shorter exposure than those on the sea-coast, and proved more intense. But we were not equally ready to receive, until so prominently brought up by these Orotava, immediately after Alta Vista, experiences, that chemical rays suffer so much more disturbance and dispersion by the atmosphere, than do the luminous ones. Here, however, was the proof; for at and near the sea-level, our photographic camera saw only a dim outline of mountain, looming through a chemical fog; where the eye, though sensible of an atmosphere be-

tween, saw all the lights and shadows of the cliff.
On the other hand, at 10,700 feet in level, while the
eye saw better than below, the camera saw so very
much better still, as quite to vie with the organ of
man.

Photo-stereograph, No. 3, shows one of these views
of the Peak, from Orotava. The Spanish houses make
an abundant appearance; so also an eruption crater,
with its double head, and outline, exquisitely rounded,
for it is entirely composed of rapilli, no solid rock with
abrupt fractures. Next appear the sky line of Mount
Tigayga, the wall of the elevation crater, with the
Peak, or eruption crater beyond. Clearly also comes
out, by its cutting against the sky, that break in the
wall of the great crater, through which so many lava
streams were seen from Alta Vista to have flowed, the
" Portillo," or " Gate of Taoro." All these things
are distinguishable in the glass picture; but none of
the bright rock features on the surface of distant
ground; no tints that indicate zones of vegetation;
and unfortunately, little of what many might consider,
the most charming parts of the landscape they behold,
viz., the distances.

Rather disappointed at the result of our long-shot
attempt, we descended from the flat roof we had been

perched on all the morning, and wandering westward, came to several magnificent specimens of " Euphorbia Canariensis." Some ten feet high and twenty broad, these bushes displayed only a host of upright prismoidal stems; as naked, as bare, as stiff, and ill-favoured, as anything one can fancy of desert growth; a creation they were of light and raw heat, salt land and no shade, or genial moisture; a product of N.E. wind, sea-surf, and broken rocks of lava.

Growing on a rough Malpays, and eminently contrasting with the black naked stones around, these plants were thrown out into imperial grandeur of their kind, by the only other specimen of herbage, in their immediate neighbourhood. For these were tender, delicate forms of some similar succulent or stem-leaved plant, but compared to the euphorbia, a gazelle to the largest elephant.

Why was such a bush ever created, said one of the curious gentlemen who had accompanied us out of Orotava, to see the photographic operation. It is useless, continued he, for any good purpose, and has active poisonous qualities. It is too succulent to burn, too soft and herbaceous to serve either carpenter or agriculturist for wood or for sticks ; and then its milky juice is a deadly virus.

COCHINEAL GATHERERS AT OROTAVA.

p. 406. 444.

Printed by A. J. Melhuish under the superintendence of James Glaisher, Esq. F.R.S.
and published by Lovell Reeve.

We mentioned that Mr. Anderson, surgeon to Captain Cook's third expedition, had a better idea of the plant, and had tried to disabuse the people of Madeira as to its lethal nature. Thereupon our informant answered, that he had heard of some one, who had visited the island, cut a piece out of a euphorbia stem, soaked it in water, and then eat it; but he, who narrated the story, added, that *he* would not do so for anything, not for all the gold and silver you could give him in the world.

A crowd of boys had come following from the town. Not imagining that we could be so absurd, as really to be interesting ourselves about the proscribed vegetable,—they wandered up the ravine, and soon finding a fig-tree in fruit, they called for us to come, for their tree was something worth the observation of sensible folks. Finding after a while that their advice was not taken, they came back again, and soon began to exhibit the popular feeling towards an unfortunate euphorbia in the neighbourhood. Shouting out terrible Spanish maledictions against it, they threw volleys of stones; screeching with elfin joy, and using worse names than ever, when from the bruised stalks, the characteristic white milky juice of euphorbias, spirted and poured out in torrents.

Persuading them at last to be somewhat more orderly, *Photo-stereograph, No.* 14, was obtained, where the white-bleeding wounds of the plant may be seen above the head of the standing boy. His companion's dress, consisting of nothing but a coarse shirt, we thought he would look more appropriate in a slumbering attitude.

Proceeding inland from this point, we entered some cactus gardens, cultivated for the growth of cochineal. In *Photo-stereographs, Nos.* 15 *and* 18, may be seen young and old specimens of this succulent plant, *opuntia tuna;* in some of its varieties approaching the euphorbia form, but ever distinguished therefrom by transparent, instead of opaque, juice exuding, the instant a leaf is pricked. The white blotches on it, are the fibrous coverings of the precious insect, cochineal; now the chief support and mainstay of the island.

The circumstances of the introduction of this new industrial resource, bear quite a providential aspect. Who would have thought in 1835, that the years of the grape vines of Teneriffe were numbered. Had it not been a vine-producing country for 300 years, and what was to prevent it going on for ever, said naturalists of non-secular progress. So when a native gentleman introduced the insect, and its appropriate cactus from Honduras in that year, his friends thought him a simpleton, and the country people destroyed his plantations at night; because they were an innovation not to be tolerated in a land of grapes.

The Government, however, happily supported the spirited improver; and though at the expense of an agrarian disturbance now and then, some cochineal and cactus were preserved in out-of-the-way parts of the island.

Time passed, and the vine disease fell on the land. The fruit withered, the plants died, starvation stared every one in the face. Orotava, so frequently visited before by Americans, anxious to exchange deals and lumber for wine,—was soon entirely deserted by that calculating people.

Then came the experiment, " cochineal" growing in the abandoned vineyards. It succeeded to admiration. The insect propagates rapidly, and its embryos spread from hand to hand. A *furor* in its favour soon took the people, and has not yet subsided. Spare land, gardens, and fields were all turned into cactus plantations. Within six months after setting out the leaves, harvesting may begin. Such

a profitable investment of the land was never made
before. An acre of the driest ground planted with
cactus was found, we were told, to yield 3 kintel, that
is 300 lbs. of cochineal; under favourable circumstances 500 lbs., worth £75 to the grower.

No wonder therefore that such enthusiasm prevailed; the men cultivating crops on a large scale in
the fields, while the woman-kind, each with some
little plot or corner near the house, were accumulating pin-money from the smallest patches. Then
adventurous persons, exploring ravines and moun-
tain-slopes; wherever they found any old cactus-
plant,—that might have long braved the sun and
the breeze in some undisturbed nook,—pinned on it,
with its own thorns, the mark of this little cochineal
beast, viz., the rag carrying its young insects. These,
very minute, are produced by the parent, in great
numbers. The few males amongst them take the
form of a gnat, live a short life, and die; leaving
the female, in look something like a bug, only white,
to perform its useful and laborious mission of se-
creting so much purple fluid. When fairly charged
with this, they are taken off the plants, placed on a
board, and baked to death in an oven, to constitute
the dried preparation of the markets.

Cochineal thrives best in the south of Teneriffe,

where the growers have two crops in the year. In the north they have only one, and are obliged to buy fresh insects every season from the south, as they cannot survive the severity of a northern winter. At one time, the natives from the south used to come and beg alms of their northern brethren; for though they planted vines, the fruit seldom came to maturity in such arid soil. Now the south has become the richer district; all owing to the cochineal; and its power of elaborating a brilliant red dye, out of the otherwise useless cactus; a plant too, we must say that for it, capable of growing and even flourishing, on far drier ground than the vine.

Mischances will occasionally take place, as that a heavy shower of rain may wash the insects off the smooth surfaced cactus-leaves; then great part of the stock is lost. Again, though liking a high temperature, and rather dry air, the creatures are excessively tender to radiation. Hence long before we made our black-bulb observations on Guajara, the natives had ascertained that the direct power of the sun increases so rapidly with height, that the poor insect, when taken from the sea-coast to a level of 3000 feet up the mountain-side, is killed by the fervid rays.

CHAPTER II.

DRACŒNA DRACO.

THAT Nature cannot be improved on, is a doctrine with some, both in art and science. There are painters who advocate simple copying to form a picture; under the sweeping and levelling conclusion, that everything that is natural, is beautiful; and there are physicians who oppose the idea of sending consumptive patients to a warm climate,—the absurdity, say they, of supposing that nature would have produced beings in any region of the earth, not adapted to that part, far better than to any other.

Yet in spite of such opinions, men who have been threatened with failing lungs in this country, expatriate themselves and become sturdy sheep-farmers in Australia; and there are, happily for art, many painters, who will not condescend to sit down before the first barn or tree they chance to come across, but wander over the whole country, noting thousands of trees and dwellings of men; comparing, selecting, combining, and reasoning on the differences found.

To the same effect has been the introduction of cochineal and its cactus into Teneriffe. Nature had previously set a plant in the island, *Dracœna Draco*, or the dragon-tree, producing a splendid scarlet in the form of gum, dignified by the old Arabian physicians with the mystic name of "dragon's blood." What need then, according to certain dogmatists, for introducing an insect and a plant from the other side of the world? Would not nature, they ask, have unerringly placed in each country, the tree best adapted to its climate and soil?

So far as the production of a marketable dye is concerned, the question is soon answered. Through means of those imported agents, the ground of Teneriffe begins to yield, within a few months of its tillage, an amount in money-value, that leaves every other article of produce far behind. But worked only with the indigenous tree, years, generations almost elapse, before any notable quantity of the dragon's blood can be collected. The elaboration of a dye, though possible to, does not seem to be the forte of, *Dracœna Draco;* and besides that, being one of the most slow-growing of trees, it can by no means

compete with the little cochineal, for the honour of
imparting blushing hues, to damsels' silk attire.

Yet for all this, we must not pass the dragon-tree
superciliously. There are many sides to its character,
and we have only to examine it further, to find its
claims to attention and admiration, on one account
or another, increase in proportion as we study. Nature, we must remember, does not test everything
on the £ s. d. principle; but has created much, of
which the only use yet found, is, to excite the
wonder, stimulate the faculties, and exalt the soul
of man.

For one who has not quitted this country, it is not
easy to realise a good idea of this dragon-tree; a
production quite typical of the flora of the lowest
zone in Teneriffe, as well as of North and South
Africa; yet so different from anything amongst ourselves, that when we open a botanical work on the
natural system, we find the only dispute is, whether
the tree,—which may attain the bulk and age of a
fine oak or elm,—is to be classed as a lily, or an
asparagus.

Had our civilization been the fabric of a more
southern *locale,* as that of Egypt or even Greece; it

is possible that the so-called natural system of botany would have been much closer to nature, than we find it, in a land; where, from one extreme of the vegetable world, viz., the dark and watery, we have to go to the other extreme, the bright and dry, and meet very diverse forms.

In the cactus-garden of *Photo-stereograph, No.* 15, with a small date-palm behind, blurring its long leaves by swaying about in the wind, may be seen two young dragon-trees. Like yuccas, or the smaller American agave—a plant that is merely a protracted annual, dying after the production of its one set of flowers—the dragoniers grow up; and no pure theory could tell the exalted future which awaits this strange plant, and makes it a precious link of connexion between many others.

Those little dragons then, in the state of pupilage as agaves, are probably about three years old. In thirty years, they will have shaken off that vestment, and will appear like young palm-trees, each with its tuft of long sword-shaped leaves at the top of a tall, white, and straight endogenous trunk.

Add thirty years more, and the definition of a palm is gone; for the stem branches above; and each branch, has now in its turn become a sort of palm-

tree, with a terminal bunch of leaves. Thirty years
more, and these branches having subdivided again,
we have a host, a republic, of palm-trees, mounted
high in the air. Each in itself, a strictly endogenous
tree; but at the point where its own individual stem
shoots forth, sending down roots which, inosculating
with each other and working their way through the
old bark of the original stem—give at last a look of
portly exogenous growth to the lower trunk.

Several of these stages in the *Dracœna's* appearance
may be seen in *Photo-stereographs Nos.* 16, 17, and
18, all taken in the garden of our friend Mr. Charles
Smith. Those in No. 16, may be considered the
best type of the adult plant under favourable circumstances; more like a zoophyte perhaps, or the penta-
crinites and lilyencrinites of the carbonaceous period,
than any living form of tree.

At this rate, what will the creature become ultimately, if it has already deviated so far from the
guise in which its career began,—does any one ask?
Let us set off to " Villa Orotava," and see; for there,
is a specimen of *Dracœna Draco,* to which most sober
naturalists attribute the age of 6000 years. They
say that it is the oldest tree on the face of the earth;
" so old," add the *catastrophic* geologists of the Gallic

school, that "it may have witnessed some of the
latest revolutions which our planet underwent, prior
to the advent of man." "The great dragonier of
Orotava" has thus occupied its site so long; that it
may even contend with the Peak itself, for being that
veritable dragon which protected the golden fruit, in
the beautiful gardens of the Hesperides.

What Laguna is to Santa Cruz, such is Villa, to
Puerto, Orotava; a place of retreat during the hot
months, above 1000 feet up the mountain, to all who
can afford the move, and all the year through to the
island aristocracy.   To add "Orotava" to either of
the names is considered a redundancy.   So for us,
living in Puerto, we had merely to inquire how to
get to Villa.

Mr. Goodall soon furnishing us with horses and
men, away we went one fine morning, first winding
out of the town eastward, and then ascending the
general mountain slope, nearly parallel with our first
day's ride when bound for Guajara.

The Peak was splendidly clear, Mount Tigayga's
steep-cleft side absolutely resplendent.   The long
valley of Taoro, stretched upwards in miles of perspective, clearly visible through its whole length,

right up to the " Portillo."   There, the side illumination of the sun brought out admirably, the jostling,
shouldering lava streams, which have rushed tumultuously out of that narrow exit from the great crater,
and poured down the slope in endless undulation.

With the sun glancing on, and gilding, the tops of
all these waves of stone, ἀνήριθμον γέλασμα, in its
earlier translation, " multitudinous dimpling" of
ocean's surface, was powerfully brought before us; a
later, and said to be a truer, interpretation, "irre-
pressible laughter " of the waves as they break on a
beach, we could by no means apply, either to the present or past state, of these once fearful streams of
molten, fiery rock.

Through that same " Portillo," now come every
winter, the torrents of water, that cut the ravines so
deep, and oblige the erection of innumerable stone
dykes, across and about all cultivated ground.
Through that same gate also, came the terrible flood
of November, 1829, still deemed incomprehensible by
all the natives.

A sudden light seemed to beam upon it in our
minds.   On the evening of our storm of September
14th, at Alta Vista, not a drop of rain fell at Orotava;
nay, even the visitors who left us at 3 P.M. that day,

came down, we were subsequently informed, quite dry. Yet two inches of rain fell on the Peak within a few hours after. Had we not been encamped there, no man would have known anything about it.

The rain appeared, not as an isolated phenomenon, but as a part of the process of autumn setting in on the heights, while summer was still prevailing on the coasts. As autumn advances, the rains must increase on the Peak, and fill its loose spongy material; so when winter itself begins, should a sudden fall of many inches take place, the ground is unable to absorb any more, and the entire flood rolls down the mountain side.

In that upper world, this rain may have been falling for twenty-four hours, and all over the vast area of the great crater, without the inhabitants of the towns, at the time enjoying their warm autumn weather, knowing anything about it. Our experience showed that they could not dream of what was preparing, until the deluge comes pouring through the "Portillo," in the unlooked-for, and improvised manner which they all ascribe to the catastrophe of November 6th, 1829.

Scarcely had I given utterance to the idea, than it was deprecated, derided, denounced. That flood, every one said, was no ordinary flood; no common

E E

rain made it,—it must have been a water-spout.
There had been no rain the previous night; and when
the good people of Puerto looked out in the morning,
their grand square, and every street, was covered
with water, foaming down from the mountain; and
threatening to wash the whole town into the sea.
What, they asked triumphantly, but a water-spout
could do that?

As our horses paced steadily upwards between
garden-walls, we could not but remark the quantity
of blackberry brambles in this land. In descending
the mountain we had found them at 3000 feet of
elevation, and they had continued down to the sea
level. In this well-cultivated neighbourhood they
probably would not have been allowed to appear, had
they not been useful in supplying the place of broken
glass on the top of the walls. So between such rows
of *chevaux-de-frise* we rode along.

At length reaching "Villa," we find the whole
town, of some 7000 inhabitants, with ancient
churches and fine public buildings, seated uneasily
on a slope so steep, that the water in the gutters
shoots downward with a rapidity almost fearful.

Plodding up one of these inclined streets for awhile,

THE "GREAT DRAGON TREE" AT THE VILLA DE OROTAVA.

p 419.

Printed by A. J. Melhuish, under the superintendence of James Glaisher, Esq. F.R.S. and published by Lovell Reeve.

and then turning to the left, we arrive at the mansion of the Marquis of Sauzal. Leave is asked, and readily granted. We pass through the house; descend a terraced platform, turn the corner, and are *vis-à-vis* with the great dragon-tree of Orotava. (*See Photo-stereograph, No.* 19.)

Proudly it raises its antique arms above everything around; but how it is hampered. An indigenous wild laurel tree is absolutely in contact on one side, and a Lombardy poplar is almost touching on the other; while there are such numerous peach-trees, oleanders, myrtles, and oranges between and all about, that there is hardly a single point from which we can get a fair view.

A rivulet of water flying along in front, the whole ground thrown, from its naturally sloping surface, into so many terraces for cultivation, the boundary line between two of them passing through the tree, and making the ground on one side five feet higher than on the other; hedges crossing in every direction, and the faint form of the summit of the Peak appearing between two tall, dark cypresses, complete the general accompaniments of the scene.

Poor old tree, whose trunk was hollow,—when Alonzo del Lugo and his *conquistadores* in 1493, esta-

blished the Spanish authority here—and turned the
bark into a chapel for holy mass, after it had served
Druidical purposes amongst Guanche tribes for ages :
how frail is it now. A storm in 1819 wrenched off
an arm; and more recently, certain Goths hacked an
immense piece out of the thin wall of hollow trunk,
for the Museum of Botany at Kew.

In place of growing larger, the tree was rapidly
collapsing, when the present intelligent proprietor,
the Marquis of Sauzal, came into possession. He
immediately interdicted a finger being laid on the
poor thing, (though always allowing strangers a
sight of it, notwithstanding they are arriving almost
every day) ; and by supplying the abstracted portion
of the trunk with masonry, he has given it renewed
strength.

Sixty feet high above the ground at its southern
foot; forty-eight and a half in circumference at that
level, 35·6 at 6 feet above, and 23·8 at 14·5 feet
above, or the place where the branches spring out
from the rapidly narrowing conical trunk,—this
Dracœna cannot compare with real monarchs of the
forest for size. And we must remember that it is
no proper tree, with woody substance; it is merely a
vegetable; an asparagus stalk, with a remarkable

power of vitality, and an equally eminent slowness of growth; it is this last indeed, not its size, which has gained it the credit of being the oldest tree in the world.

Let us take note of the chief characteristics. First, the immense uprearing of long naked root-like branches; and the pyramidal outline of the trunk. The leafage makes no very sensible appearance; there is the typical tuft at the end of each branch or rather stem; but the miniature palm-trees have been growing for ages without bifurcation; extending only in length, nothing in breadth. At the point of junction of two or more, a thickening of the lower branch begins, and occasionally may be seen one or two withered radicles, hanging loose; for they have failed to enter the bark, and work their way down to the ground.

So many of them, however, have done this; that, while the simple stems are smooth, or marked only by shallow, transverse indentations of foot-stalks of past leaves,—the compound stems are deeply corrugated longitudinally; and the trunk, more markedly still; with an evident tendency in every wrinkle to divide continually as it descends.

When once a stem has branched, its life seems to have departed, being replaced by the lives of the several young trees of its kind, left growing on its summit; and whose roots, entering the bark and encasing the old stem on every side, conceal its slowly withering corpse from the light of day. Ages pass by, the young trees after flourishing, die in their turn; each producing two or more new ones mounted on their summits; and altogether presenting such a surface to the wind, that the base of the original tree would never be able, unassisted, to support the strain.

See, however, the admirable provision of nature. The inosculating roots which had decorously concealed the death of their parent stem; feeling the requirements of the growing family above, expand their circle of support below; the trunk that had been cylindrical, becomes a broad based cone. An opening is made on one side; we look in, and find a mere hollow. In the centre of that void, once stood the original tree: it is gone now, as completely as any of the early progenitors of annuals growing in our gardens.

Hence some explanation of the hollow interior of the great dragon-tree; it is a physiological necessity.

But hence also a considerable limitation in the age of those parts which are still leafing and flowering, viz., the extremities of each long, thin branch. Supposing 6000 years to have elapsed since the original plant first came above the surface of the ground, that period must be divided by nearly the number of times that the tree has branched.

Yet as the successive generations must have followed each other on the self-same spot, this conclusion will not touch the dazzling theory of an eloquent author, setting forth that Dracœna Draco does not belong to the Canaries nor to Africa; but to India; and that the Guanches must therefore have had commercial relations with that part of the world more than 5000 years ago.

An attempt has been made to disprove such idea, by showing that the dragon-tree is indigenous to Teneriffe, on the ground of dragoniers having been found in out-of-the-way valleys there, and on rocks inaccessible to man. But unless those specimens be older than the Orotava patriarch, why should not winds or birds have carried the seed from a tree, imported into a country with favourable soil and climate.

Of the suitability of these two, there is no doubt;

and if a great man who was never in Africa has
simply asserted that the tree is not there; where is
the proof, we are entitled to ask. The country of the
Moors, and much besides has not yet been botani-
cally examined: and Morocco has long traded in
dragon's blood, that could not well have been procured either from the Atlantic islands or from India.

Probabilities then, in place of being against, are
rather in favour of *Dracœna Draco* being indigenous
to N. Africa; and if so, most likely in the Canaries
as well; while in the similar latitude of S. Africa,
among the Khamies-berg range of the western coast,
is a well known tree of the same family, the Koker-
boom.

A drawing of this made for me some years ago, by
my friend Mr. Charles Bell, surveyor-general of the
Cape Colony, is precisely an old dragonier in its
mode and form of growth; the pyramidal trunk,
the single stem branches, and the terminal tufts of
leaves.

Showing him, on the other hand, a few weeks since,
without any remark, *Photo-stereograph*, *No.* 19, he
immediately exclaimed that it was a Koker-boom.
But what did he say when Professor M'Gillivray's
view of the great dragon-tree was laid on the table?

What could any one say, on seeing a huge *elm* tree, with a superabundance of small leafed foliage, a height of 150 feet, as measured by the man going up the ladder, and the position solitary, in a nearly level country.

No one copying nature direct, could have erred so widely; but then our Aberdeen professor had copied Baron Humboldt; so we looked up his view in the celebrated "Atlas Pittoresque;" and found that the tree was there only ninety-five feet high, not with elm foliage, but rather that of a sycamore, in the Italian style of line engraving; the trunk was smooth, sleek, and seemed to contain 20,000 cubic feet of solid timber; and the *locale*, showed us a very flat alluvial meadow.

Baron Humboldt's view again, not being taken from the tree itself, but from a drawing by M. Marchais, and that from a sketch by M. Ozone; we applied to the hydrographical department of France, where Ozone's papers are still preserved, and a certified copy was forwarded through the courtesy of Admiral Mathieu.

Here the tree is reduced to its correct height; and though the trunk has not the characteristic pyramidal

slope, and the foliage is still too abundant, yet one
can believe it to have been taken from nature; by
some sketcher, in a great hurry, and with his imagi-
nation and style of drawing based on European types
of plants.

With these three views before us, it is instructive,
as connected with the language of drawing, to trace
the gradual growth of error and conventionality, as
man copies from man.   Errors are always copied, and
magnified as they go; seldom are excellences repro-
duced.   After a few removes, the alleged portrait of
nature, is only a caricature of the idiosyncrasies of the
first artist.

Never was the debt that mankind owe to the
inventors and organizers of photography, Talbot,
Daguerre, Herschel, and Archer, more apparent than
in the case of the dragon-tree.   Artists, landing
for a few hours from a ship, were appalled at the
tangled mass of vegetation about the old dragonier,
and made a sort of ideal tree, on a bare level surface.
Nature, on the other hand, awed by nothing she has
made, takes on the collodion plate, the whole scene,
with all its foreshortenings, all its groupings, as in-
stantaneously as a flat wall.

"That cannot be long enough," said one of our

TRUNK OF THE GREAT DRAGON TREE

p. 427.

Printed by A. J. Melhuish under the superintendence of James Glaisher, Esq. F.R.S.
and published by Lovell Reeve.

companions, as we tried a half second of exposure :
but we showed him presently the hollow trunk, the
wrinkled bark, the gardener's scaffolding, which has
passed into the fiction of the ladder and man; the
long branches, the sword-shaped terminal leaves,
hedges and terraced land, distant trees and still more
distant hills.

One more view we took, *Photo-stereograph, No.* 20,
exhibiting the trunk from the east, where its cala-
mities are least apparent.   On the masonry filling up
the northern side, is standing our interpreter, the
Vice-Consul's son, to afford a human scale of measure;
his unfortunate hand, so bruised at Alta Vista, still
in a sling.

CHAPTER III.

ADIEU.

SEPTEMBER 26th saw us riding from Orotava to Santa Cruz.

During the first part of the journey, while still in that depressed or fallen-in region of the valley of Taoro; gardens were thick and close. Of their contents, none were perhaps so admirable as the double oleander; think of a rose-bush at least twenty feet high, and in a perfect explosion of flowers. Peaches were fine, as trees, but the fruit, strangely small; though numerous as the leaves. Date-palms, *Phœnix Dactylifera*, were not unfrequent; and after ascending the hill surface beyond the valley, villages were passed with almost groves of them. Yet none could be called well-grown. Stunted and stumpy they rather were, more like tree-ferns than palms; and whenever fruit appeared, it seemed dropping off immaturely. Whether the soil or climate be at fault, others must say. The latitude could not be, for in Egypt and Syria, several degrees

north of Canary, date-palms wave on stems as tall
and graceful as any in the world.

These countries, however, have no Trade-winds
blowing on them every day, as in Teneriffe; not only
bringing cold air from the Poles, but producing such
mists through the greater part of summer, that the
chief portion of solar heat, due to latitude, is reflected
back to the blue sky, from a brilliant upper surface of
an almost permanent sea of clouds.

Under such discouragement, vegetation along the
northern coast of the island, is simply not so fine as
it might be; with certain exceptions, there is more
of the desert, than of the " scenery of plants."
These, often most curious, have been admirably described, with all scientific minutiæ, by Barker-Webb,
and Berthelot, in their unrivalled volumes.

The industry of the inhabitants was perhaps the
chief source of interest that rose before us. They
have certainly worked out the garden system of
cultivation admirably; and enabled the land to support astonishing numbers. Hills terraced all the
way up their sides, and corn growing by handfuls
in corners amongst rocks, where the only moisture
that can visit it, is sea-fog,—form a sight to im-
prove the inhabitants of our Hebrides, where so

many hills and valleys, are equally untouched, un-
altered, and unimproved by man.

In spite of deserts of lava and pumice, and of
regions elevated above the clouds, Teneriffe on the
average maintains 106 individuals to every square
mile: our colony of the Cape of Good Hope, about
one to every ten miles.

After traversing half the distance between Orotava
and Santa Cruz, riding over mule paths, we came on
the new road that Government is constructing to
connect the two towns; a broad and admirable
carriage way. An immense impulse will this give
to improved ideas; in a land, for a colony, reverend
with age, boasting of an indigenous aristocracy,
marquises and counts of 300 years' standing, yet
until recently without a wheeled vehicle for country
transport.

Already an omnibus is laid on to carry passengers
the length of the finished road from the capital; and
we heard of three other carriages; though one of
them was *hors de combat* indefinitely, from the difficulty of getting experienced blacksmiths to repair a
broken part.

Somewhat late, in truth, have the Spaniards been here
in forming roads; but having begun, they are doing the

work well. A new broom sweeps clean; and thus, while you may preach for ever to Scottish macadamizers, not to injure the feet of all the horses in their country, by covering roads with naked, angular, cutting stones, leaving it to unfortunate animals with delicate feet to stamp the mass down, and wear it smooth,—a more humane, as well as more perfect, mechanical plan is adopted in Teneriffe. Before the public is admitted to a new portion of the road, a gigantic roller, loaded with tons of lava, is dragged backwards and forwards over the materials of macadamization. As these are gradually flattened down, triturated siftings are added, and the surface is speedily brought to a condition, that you might shoot marbles along it.

With a wind blowing behind us from the north, we entered the elevated country about Laguna, and soon were in the midst of driving rain. So had it been when we came through this same tract, in an opposite direction, last July; and so we concluded that it must have been, from our observations on the Peak, nearly every day since. Probably not quite so bad as that; but the numerous plants of *Sempervivum urbicum*, growing into magnificent heads on the roofs of houses in the extensive city of Laguna, speak as strongly of much wet and little sun, as do the grasses so plentiful in fields around.

The thickest part of the cloud was passed through, before reaching the ancient capital; where we found two winds blowing in, from the east and south-west, respectively. The conflict of these currents with the northern, over the tract between Laguna and Tacoronte seems a main cause of the deposition of so much rain there; and appears chronic. Von Buch has noticed the strange circumstance of two windmills in this neighbourhood, with their sails directed to opposite quarters, and each with a fair wind whenever he passed.

A ride of this sort up into the lower parts of the cloud, hereabouts hanging eternally at almost 2000 feet above the sea, must have given travellers a fearful idea of dismally cold and watery regions, awaiting that persevering explorer, who should venture further in the same direction. The conclusion would not be a Canarian one only; for all the world round, are not clouds generally raised some thousands of feet over the earth; and do not rains come down to us from above.

Ascending mountains at the Cape of Good Hope,

for instance, from dry air below—we enter a cold wet
mist on the heights, with all its botanical consequences.
Thus, on Hottentot Hollands Sneeuw Kop, from
heaths and proteas, flowering evergreen shrubs with
aromatic odour, and hard woody stems below—we
arrive by degrees at grass and reeds on the top.
Everywhere the woody forms of the lower regions,
pass into grassy, in those above. The flats at the
foot of Table Mountain are composed of arid white
sand, with here and there a large bush of *Protea
mellifera*, carrying huge pink and white cup-shaped
flowers; the flats on the top of it, are black and
boggy, percolated by rivulets of water, and closely
covered with short reedy growths.

Similar phenomena were observed in other countries
also, and at once the generalization was leapt at,
" the summits of mountains are misty and wet, in
direct proportion to their height." On this belief, a
scheme of plant-zones was made out for the Peak of
Teneriffe, on the principle of the smaller hills actually
observed; and beginning with vines below, ended
with a region of grasses at the top.

With such detail before him,—as the result of
observation, and with the specific note by one author, of the " compact short-swarded turf being so

F F

slippery," as often to endanger his falling, between
Alta Vista and Estancia de los Ingleses at an altitude
of 10,000 feet,—the late Professor Daniell was a bold,
but clear-sighted, man, when he published his con-
clusions of *dryness* existing above the clouds, usually
floating at less than half the level, of the just-
mentioned cumulative plant-proofs of moisture.

From the Alpine journeys of the scientific Saussure,
some instrumental indications of what had thus
come to be suspected, were first obtained. Next Mr.
Green, the well-known aeronautist, being furnished
by Daniell with one of his admirable hygrometers,
brought down from a balloon voyage, undoubted
proofs of the new theory. General Sabine followed
with observations on the Blue Mountains of Jamaica;
and Mr. Welsh's four recent ascents by balloon, have
now unalterably established the fact, that from the
surface of the earth up to the level of the first,
Newton's "grosser," clouds—moisture evidently increases; but above that level—suddenly and greatly
decreases, barring exceptional cases, to more than
African dryness.

So much for the atmosphere generally; but what
of the particular instance of Teneriffe. As regards
numerical observation, the dew-point depression from

Orotava up to 3000 feet was, with us, seldom more than 8° to 10°; at that height (the full measure of many of the wet-topped South-African bergen), if the clouds were dense, there was saturation of moisture; but above it, the depression or dryness increased, not quite so suddenly as indicated by Daniell, but rapidly; until, at 8900 feet of altitude, a depression of 56° was attained at a temperature of 55°. At the same time, these upper regions of the mountain presented everywhere a sterile aspect; any plant that was seen thereabout, invariably proved a woody bush, more like the growth at the bottom, than the top, of Table Mountain; and at the place, 10,000 feet high, of the "compact short-swarded turf" already noticed, we found only a rolling slope of clinkery ashes, pumice, and occasional fragments of obsidian.

Grass then, was by no means discoverable towards the upper part of the Peak; and we had to descend more nearly to the bottom of it, or to 3000 feet of altitude, before that peculiar class of plants could be conspicuously met with. In other words, we had to enter the lower stratum of Trade-wind cloud.

Looked at timidly by some, hastily by others, from fertile valleys below, the upper parts of high mountains have been too often sadly misunderstood; and have had such untoward climates attributed to them, that, although the idea of eliminating atmospheric tremors from telescopic vision, by their means, was promulgated by so great a man as Sir Isaac Newton, —the scheme fell dead on the world, and no serious attempt to try it was ever made, until this mission to Teneriffe in 1856.

In physics as well as in mathematics, some remarkable intuitions were gained by that wonderful mastermind; and not the least admirable among them, though the last to have the proof by experiment applied to it, is that which bears on the probability of "*a serene and quiet air, pre-eminently fit for astronomical observation, existing on the tops of the highest mountains above the grosser clouds.*" The prescience exhibited herein, is illustrated all the more powerfully, when we find the most popular physical teacher of the present day, both describing the Peak of Teneriffe as constantly enveloped in cloud, and specially recommending low, *versus* high, positions, for telescopes directed to the heavenly bodies; because, according to him, "mountains have a misty and variable climate;" and, "the elevated strata of the atmosphere, when they envelope the ridges of mountains, undergo rapid changes in their transparency."

Pascal's suggestion of a mountain ascent, barometer in hand, produced an epoch in the science of
pneumatics. Newton's idea of going up with a
telescope, may be of more signal advantage still, to
astronomy,—if it be energetically carried out in
practice.

From Laguna the ground descends rapidly towards
the sea.   Leaving the new carriage-road on the west,
we rode down by the grand mule path, as important
a work in its way, as the road; being nearly as broad,
paved through its entire length, bordered with parapet
walls, garnished with landing-places, and carried over
and alongside remarkable torrent-beds, showing worn,
but not grooved, basaltic rocks, at a giddy depth
below.

Arrived in Santa Cruz, we are met by our friend
Mr. Hamilton. Captain Corke is also there, and
undertakes to get the yacht under weigh, while we
are examining a tide-guage recently erected for us by
our friend, and the engineer of the Mole, Don
Francisco Aguilar.   It is admirably constructed; but
who is to observe it?   The Spaniard's enthusiasm
for a piece of useful science is equal to any one's; and
he readily volunteers to do all that will be required,

though the very first step has been already intimated, as nothing less than—observations at five minute intervals, for three days and nights consecutively.

From first to last we owe our sincerest thanks to all the inhabitants of Teneriffe, with whom we have had to do, from the Captain-General downwards. Coming as perfect strangers, yet for scientific purposes, we have been received and aided rather as countrymen.

What can we say more. The Titania awaits us, with her sails already spread; and she must be round the island, before Orotava, sometime the next day, Saturday, or remain there idly until Monday.

The winds through the night are light and variable; but somehow, the sharply entering iron hull does really " walk the waters like a thing of life." At six o'clock, on Saturday morning, some one reports to Mr. Goodall in Orotava, that the yacht is in the offing. He declares that such a thing is impossible, ascends his belvedere, telescope in hand, but immediately comes down again, confessing that it is the yacht; and he arouses porters and boatmen to bring off our instruments and stores.

With us at sea, the wind had died away to a calm; the sky was clear, the sun bright, the swell had decreased to a ripple, and the Peak of Teyde rose before us in all its grandeur, with hardly a portion of cloud about it.

Immediately over the line of the dancing wave tops, were the buildings of Puerto Orotava; beyond, the two rapilli craters; and then the broad white surface of Villa, where the telescope even distinguished the mansion of the Marquis of Sauzal, and the dark, peculiar form of the great dragon-tree. But up above all this; above the long valley of Taoro, above the Portillo and Mount Tigayga, arose more magnificently than anything else, the Peak itself, the " crater or cone of eruption." In Orotava, it is evidently always foreshortened by the walls of the " elevation crater ;" but six miles out at sea, we saw it in full proportion, " one mountain upon another," as Von Buch has well described it.

A black speck is now seen on the waves towards Orotava; then is lost behind a ridge of the swell, then is seen again; there are two; and before much longer, we distinguish our trusty friend, the acting Vice-Consul, Andrew Goodall, coming off with all our packages.

He is soon on board, and brings with him a cluster of bananas, so large as to need almost to be carried on a pole between two sturdy coolies. We would willingly prolong the meeting, but he is far from shore, and must get back there safely, before a wind springs up.

From the stern of the yacht, now directing her bows towards England, as we watch the vanishing boat of this excellent, and hard-working official—we think of his innumerable services to make our long residence on the mountain, effective for science and agreeable to ourselves.

As evening advances, Orotava is lost, and the cinder hills, and the Villa, below the blue edge of ocean. Clouds too form all along the 3000 foot level; but above all this, is still seen the great Peak, standing on the vast plateau of the elevation crater; raised high above all the turmoils of this lower world, into the calm grandeur of height.

At length, when night closes in, and our last view of the Peak leaves it still high in mid-air,—we wonder how long the learned world will delay to occupy a station, that promises so well, for greatly advancing the most sublime of all the sciences.

INDEX.

THE END.